Wetland PLANTS OF Ontario

Wetland PLANTS OF Ontario

WRITTEN BY
Steven G. Newmaster
Allan G. Harris
Linda J. Kershaw

CONTRIBUTIONS BY
Robert F. Foster
Gerald D. Racey

LONE
PINE

The Publisher: *Lone Pine Publishing*

10145-81 Avenue
Edmonton, AB T6E 1W9
Canada

1901 Raymond Ave. SW, Suite C
Renton, WA 98055
USA

Canadian Cataloguing in Publication Data

Newmaster, Steven G., 1967–
  Wetland plants of Ontario

  Includes bibliographical references and index.
  ISBN 1-55105-059-5

  1. Wetland plants—Ontario—Identification. 2. Wetland plants—
Ontario. I. Harris, Allan G., 1960— II. Kershaw, Linda J., 1951–
III. Title.
QK203.O5N48 1997   581.92'9713   C96-910496-0

Senior Editor: Nancy Foulds
Editorial: Roland Lines, Jennifer Keane
Design: Bruce Timothy Keith
Cover Design: Carol S. Dragich
Cover Photo: Harold Kish
Layout & Production: Carol S. Dragich, Gregory Brown
Separations: Elite Lithographers Co. Ltd., Edmonton, Alberta, Canada
Photographs and illustrations in this book are used with the generous permission of their copyright holders. Page 234 of this book, which lists photograph and illustration credits, constitutes an extension of this copyright page. Illustrations from *Vascular Plants of the Pacific Northwest* by Hitchcock et al. are reprinted with the permission of the publishers, the University of Washington Press. © 1961, 1964, 1969 by the University of Washington Press.

Funding for the development and printing of this publication has been made available in part through the Northern Ontario Development Agreement, Northern Forestry Program (NODA).

The publisher gratefully acknowledges the assistance of Alberta Community Development and the Department of Canadian Heritage.

CONTENTS

DEDICATION

This book is dedicated to all naturalists, young and old, who have an insatiable desire to explore the swamps, bogs and fens of our landscape.

ACKNOWLEDGEMENTS

Special thanks to Michael Oldham for his thorough review of the text. Bill Crins, George Barron, Bruce Ford, Ago Lehela, Steve Taylor, Dale Vitt, Howard Crum, Joan Riemer, Ruth Berzel and Sylvie Mauser assisted in researching and editing the text. Brenda Chambers, Bill Crins and Don Cuddy offered suggestions for the species list.

The Sleeping Giant Park Library and John B. Ridley Research Library, Quetico Park provided access to their slide collections. Thanks to Adolf Ceska, Anna Roberts, Brenda Chambers, Frank Boas, Mike Bryan, Sue Bryan, Bill Crins, D.R. Gunn, Derek Johnson, Gerry Racey, Donald Sutherland, Emma Thurley, Cathy Paroschy Harris, Jennifer Line, Harold Kish, Karen Legasy, Robert Norton, Mike Oldham, Jim Pojar, Richard Sims, Shan Walshe, Trygve Steen, Wasyl Bakowsky, the Thunder Bay Field Naturalists and the Ontario Ministry of Natural Resources for providing slides.

Thanks for illustrations to J. Bowles, Daniel Klassen, Shayna LaBelle-Beadman, Linda Kershaw, Annalee McColm, Erika North, Emma Thurley and the University of Washington Press. Line drawings from Britton and Brown (1913) have also been used.

INTRODUCTION

Wetlands—permanently or seasonally waterlogged areas (including lakes, rivers, marshes, swamps, bogs and fens)—are an integral part of the Ontario landscape. One of their most important functions is to slow and sustain local water flow. On a global scale, northern wetlands may hold huge accumulations of carbon, because cold temperatures and lack of oxygen under extensive peat moss mats slow the decomposition of organic matter.

Wetlands also contribute significantly to the diversity of life and ecosystems by providing habitat for plants, mammals, birds, reptiles, amphibians, fish and countless invertebrates. Moose feed heavily on extensive beds of pondweed, pond lily and arrowhead in quiet waters of lakes and ponds. Loons build their nests on floating, lakeshore vegetation, and in the autumn, waterfowl gorge on wild rice before beginning their long migration south. Giant muskie lurk under pond lilies in secluded bays, and young smallmouth bass hide from predators under the cover of floating plants.

A wetland is covered by water or saturated by water for at least part of the year. Its soils and vegetation are influenced by the wet environment. Wetland development is related to climate and physiography, and to glacial and geological history.

In western Ontario, near Rainy River, the relatively flat, low, clay deposits left by glacial Lake Agassiz support extensive peatlands, while the neighbouring shores of Lake of the Woods exhibit sandy beaches, dunes and shallow, open water with aquatic vegetation.

In the colder, moister, clay belt of northeastern Ontario, flat terrain and deep, poorly drained, cold soils support extensive, treed peatlands. Plants of these forested wetlands (e.g., black spruce, labrador tea and leatherleaf) are well adapted to the harsh climate of this area, at the edge of the the world's largest peatland, the Hudson Bay lowlands. Poor drainage and cold temperatures result in the accumulation of peat (undecomposed plant material).

The rugged, battered shores of Lake Superior and Lake Huron support their own scattered, resilient wetlands. The wave energy of seasonal storms and the scouring of ice leave only the hardiest plant species. Tufted bulrush and shrubby cinquefoil persist under these harsh conditions.

The shallow soils and rugged terrain of the Canadian Shield support many lakes. Wetland plant communities occupy the shallow bays and margins of lakes, streams and glacial kettle holes. Beavers control the evolution of many wetlands. Streams can be transformed into beaver ponds, develop into marshes and drain to form beaver meadows in as little as 10 to 20 years, or over much longer periods. Such dynamic wetlands change visibly from year to year. In raised bogs, on the other hand, plant communities may remain relatively stable for decades and possibly centuries.

Appreciation of wetland composition and structure requires an understanding of wetland plant communities. What plants are present? How are they arranged within the wetland, and why? Some wetland plants are found in many wetlands, under a wide variety of physical and chemical conditions. Others live only in very specific environments, and they can be used as *indicator species* for their habitat type.

Wetlands differ in appearance and in biological function. The *Canadian Wetland Classification* system describes five general wetland types: bog, fen, marsh, swamp and shallow open water. One of the most comprehensive discussions of Canadian wetlands is found in *Wetlands of Canada* by the National Wetlands Working Group (1988). Wetlands can be further classified using physiognomy (the general appearance of the vegetation). The physiognomic system for naming wetlands distinguishes between shallow open water, marshes and meadow marshes; open and treed fens; open and treed bogs; and hardwood, conifer and thicket swamps. The species composition of wetland types varies greatly across Ontario. You should be able to describe most of these wetland communities using the plants included in this guide.

Most of the plant species described in this guide grow in wetlands across eastern North America, and the ranges of many extend west to British Columbia and Alaska, but this guide focuses on common and indicator wetland plants from central and northern Ontario.

Forest Regions of Ontario and Surrounding Areas

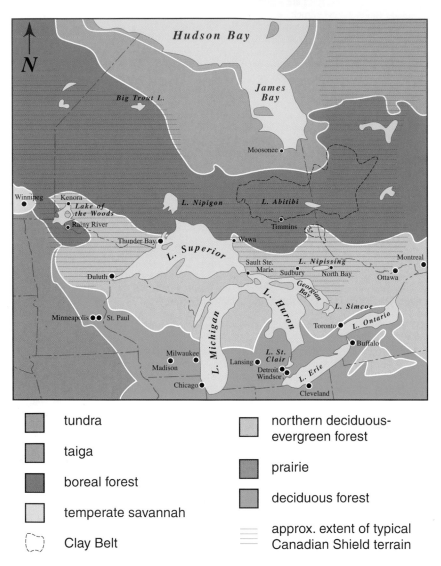

- tundra
- taiga
- boreal forest
- temperate savannah
- Clay Belt
- northern deciduous-evergreen forest
- prairie
- deciduous forest
- approx. extent of typical Canadian Shield terrain

Bogs are characterized by low nutrient availability. Organic deposits are usually over 40 cm deep, often derived from poorly decomposed peat mosses. As peat accumulation gradually raises the surface of a bog above water level, rainfall and dustfall become the only sources of nutrients. In all cases, bogs are isolated from external sources of mineralized water. This is the case in an isolated bedrock basins on the Canadian Shield. Bogs can be treed or open, but they are always acidic (they have a low pH), with ground cover dominated by peat moss.

A typical low shrub bog near Rainy River, Ontario, characterized by a layer of ericaceous shrubs (including bog rosemary, bog laurel and leatherleaf) over a carpet of peat moss. Dense cottongrass is also very common.

This graminoid-rich, treed bog has a carpet of peat moss, sedges and slow-growing black spruce (stunted by a lack of nutrients and by low temperatures in the rooting zone). Low shrubs grow on drier hummocks.

Fens are fed by nutrients from slow-moving ground water, and they are less acidic than bogs. Fens often have floating mats of vegetation that rise and fall with the underlying water. Nutrient availability varies with the flow rate and mineral content of the water. Fens support a wide array of plant communities, including treed and open wetland sites, and their ground cover is dominated by brown mosses or peat mosses.

Graminoid fens are occupied by sedges and grasses over a spongy, often buoyant mat of vegetation. They range from relatively poor to rich in nutrients.

This low shrub fen has abundant low shrubs (including leatherleaf, bog willow and dwarf birch) and scattered, stunted tamarack and black spruce. Sedges, grasses and herbs cover the moderately to poorly decomposed peat.

This shore fen, adjacent to a meandering stream, has a floating mat of poorly decomposed peat moss held together by roots and rhizomes. Abundant low shrubs and sedges overlay the moderately to poorly decomposed peat.

Shallow open waters include areas in lakes or rivers with water less than 2 m deep and with scattered, emergent vegetation covering less than 25% of the surface. These areas often have floating or rooted aquatic plants. Substrates are either mineral or organic.

This shallow open water has abundant aquatic plants with floating leaves (pondweeds and water lilies) or with completely submerged plants (coontail).

9

Swamps are dominated by trees or shrub thickets. They are generally characterized by standing or slow-moving water that may cover the land seasonally or for long periods of time. The substrate can be mineral or organic. Occasionally, the water table drops, creating a well-oxygenated rooting zone for plants. Most swamps are nutrient rich.

A typical thicket swamp with clumped, tall shrubs (including willows and speckled alder) and a substrate of well- to poorly decomposed peat and muck. Periodic flooding and the presence of hummocks, channels and pools makes walking difficult. Nutrient availability is high to poor.

Hardwood swamps usually have very high nutrient availability, and they support black ash, elm, yellow birch and red maple trees. They vary greatly in appearance and species composition, and they often support large numbers of shrub and herb species.

Conifer swamps may have high nutrient availability, and they support large, relatively vigorous black spruce, tamarack or white cedar trees, over an understorey rich in grass, sedge and herb species. The ground surface is uneven because of hummocks and depressions.

Marshes have saturated rooting zones for most of the growing season. They are characterized by emergent vegetation and relatively high oxygen levels in the rooting zone. The vegetation often shows distinct zonation with changes in water depth and exposure to wave action.

This meadow marsh has continuous grass and sedge cover on a mineral or muck soil that is seasonally flooded. Tussocks are common, and nutrient availability is moderate to high.

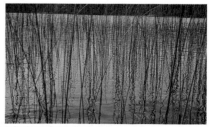

Marshes are characterized by emergent plants, such as cattails, bulrushes and arrowheads. Submerged plants are often present in standing or slow-moving water, such as stream and lake margins. The substrate is often well-decomposed peat and muck, but it may be mineral in areas with high wave or current energy.

How to Use This Guide

It can be quite confusing to sort through various plant guides to find the name of the plant in front of you. This field guide is designed to help you identify plants in the field. Unfortunately, field identifications are not always 100% accurate, and if there is any doubt (and the plant is not rare in the area), you may want to make collections that can be checked later.

Plant Names

Plants often have many common names in many different languages, but each species has only one scientific name. The system is quite simple and it is accepted worldwide. Each plant has a **species name** made up of two words. The first word is the **genus** (noun); the second word is the **specific epithet** (adjective). An example of this is *Thuja occidentalis*: genus *Thuja*, species *occidentalis*. The abbreviation 'sp.' (species) may be used if a species is not known. For example, '*Potamogeton* sp.' refers to a plant in the genus *Potamogeton* that could not be identified to species.

Scientific names are Latinized, but they may be derived from words of other languages (e.g., Greek, Arabic, French) or from the names of people or places. They often refer to a unique feature of the plant, such as habitat, growth form, or colour. These names can be used as a tool to help you remember the plant.

The nomenclature (system of names) used in this book follows *The Ontario Plant List* (Newmaster et al. 1996).

How to Identify Plants

Many of the terms used to describe plants in this guide may be new to you. The **illustrations** and **glossary** will help you understand the keys and plant descriptions. **Pictorial keys** illustrate diagnostic features that can be readily observed in the field. Keys include only those species described in this book. This guide will help you to identify most of the plants you will encounter in wetlands in this region, but all wetland plants could not be treated, and **the plant you are trying to identify may not be described here**. If this is the case, you will need to refer to some of the more detailed reference books listed in the references.

The guide is divided into six sections: woody plants (trees and shrubs), herbs, grass-like plants, aquatics, ferns and allies, and bryophytes (mosses and liverworts). When identifying a plant, first determine to which of the six sections it belongs.

Identifying a plant is like solving a mystery. The clues are the distinctive features of the plant, including plant form, leaf shape, flower colour and so on. The more clues or plant features you can recognize, the greater your chances of identifying the plant. Don't be overwhelmed by technical terms and scientific names. Learn a few at a time.

This guide avoids technical terms whenever possible, but there are some basic terms that you need to know to identify plants. Many are defined in the glossary and common terms are illustrated inside the cover or in group introductions.

Using field guides, floras and other publications, you can apply your clues to determine a plant name. Herbariums (collections of preserved and labelled plant specimens) at universities, museums and government research centres are also an excellent resource, where you can compare your sample with specimens that have been identified by professional botanists. Herbariums usually have well-stocked botanical libraries, and they are open to people who want to learn about plants and plant identification.

Caution

We are NOT RECOMMENDING the use of any of these plants for food, medicinal or healing purposes. Many plants in our region are poisonous or harmful if consumed or used externally. The information about food and medicinal uses of plants is provided for interest's sake and historical value only. The information has been compiled from other books and its accuracy has not been tested. Harvesting wild plants can also be harmful to plant populations and the wetland environment.

WOODY PLANTS

This is not an exhaustive treatment of the woody plants of this region, but it does include the common species found in the wetlands. The two major groups in this section are trees and shrubs.

TREES are large, perennial, woody plants with a main trunk and branches that form a crown. At maturity, trees are typically more than 10 m tall. This definition does not take into consideration dwarf or stunted trees, such as the ancient black spruce trees of northern bogs which can be only 1 m tall. Only 12 common wetland trees are included, but they cover a large portion of Ontario's wetland habitat.

SHRUBS comprise a large, diverse group of woody plants that are smaller than trees (usually less than 10 m tall), and usually have more than one main stem. Many shrubs take on different growth forms in different habitats, and this can make routine identifications more difficult. The willows (*Salix* spp.) can be especially difficult to identify, and it is often wise to take a voucher specimen to verify your identification.

There are no keys in this section, but species are grouped together in the following categories, based on leaf shape and arrangement, represented by the following icons:

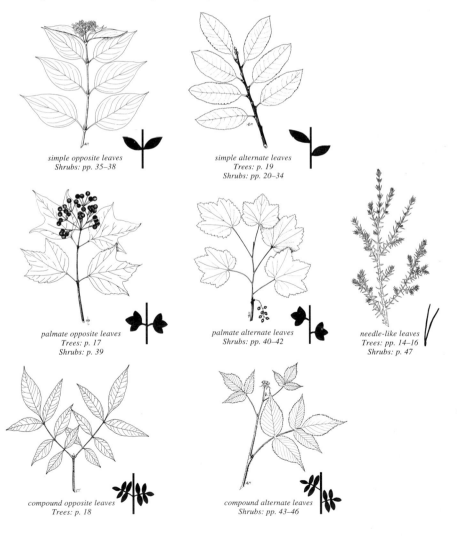

simple opposite leaves
Shrubs: pp. 35–38

simple alternate leaves
Trees: p. 19
Shrubs: pp. 20–34

palmate opposite leaves
Trees: p. 17
Shrubs: p. 39

palmate alternate leaves
Shrubs: pp. 40–42

needle-like leaves
Trees: pp. 14–16
Shrubs: p. 47

compound opposite leaves
Trees: p. 18

compound alternate leaves
Shrubs: pp. 43–46

TAMARACK, EASTERN LARCH
MÉLÈZE LARICIN • *Larix laricina*

GENERAL: Deciduous coniferous tree, up to 20 m tall; bark smooth when young, scaly at maturity; **leaf-bearing twigs short and spiky**; buds rounded, small, red.

LEAVES: Soft, green needles, 10–25 mm long, **in clusters of 10–25**, turn bright **yellow and fall off** in autumn.

CONES: Rounded, 1–2 cm in diameter, on short stalks, unisexual, with male and female cones on same tree; appear in early spring, open in autumn.

FRUITS: Winged seeds, 8–10 mm long, wedge-shaped, drop freely from female cones.

WHERE FOUND: In fens, swamps and occasionally bogs; from Newfoundland to Alaska, south to central British Columbia, Minnesota and New Jersey.

NOTES: Tamarack leaves and buds are occasionally eaten by grouse. Hares, porcupines and squirrels feed on the bark and seeds. • The Ojibway called tamarack *mushkeegwautik*. They used the bark in making wigwams, sewed with the young roots and wove bags from the roots. Tamarack could be called 'the local boat repair station,' because the Ojibway used the sap as a waterproof glue for mending boats and used the roots to sew boat parts together. In Europe, tamarack was once an important timber for joining ribs to deck timbers on ships. The Royal Navy also used it as a substitute for compass oak. The tannin-rich bark has been used to tan leather and to make turpentine. • Teas were made by boiling the roots or the bark. The bark tea was used to cleanse wounds and sores and to treat bronchitis.

TREES

BLACK SPRUCE
ÉPINETTE NOIRE • *Picea mariana*

GENERAL: Coniferous, evergreen tree, often with a **distinctive, club-like crown**, often shrub-like and less than 10 m tall in bogs and fens; **young twigs densely covered with short reddish hairs**; buds at stem tips often hidden under long, hair-like scales that extend beyond them.

LEAVES: Stiff needles, 6–18 mm long, **square**, blunt-tipped, pointing upwards.

CONES: Unisexual, with male and female cones on same tree; female cones egg-shaped, 1.5–3.5 cm long, with spirally arranged scales; appear in early spring and remain on the tree for several years, open in winter; male cones tiny, short-lived, deciduous.

FRUITS: Winged seeds, in axils of scales of female cones.

WHERE FOUND: In swamps, bogs, poor fens and moist or dry upland forests; from Newfoundland to Alaska, south to Saskatchewan, Minnesota and Pennsylvania.

NOTES: **White spruce** (*P. glauca*) could be confused with black spruce, but it has hairless branches, short hair-like scales around the bud at the stem tip and longer cones. It often grows in swamps and on riverbanks, but usually it prefers drier, warmer sites than those occupied by black spruce. White spruce is found from Newfoundland to Alaska, south to Wyoming, Minnesota and Maine. • **Balsam fir** (*Abies balsamea*) is easily distinguished from black spruce by its flat (rather than 4-sided) needles, which have 2 white lines of stomata on the underside, and by its large (5–10 cm long), erect cones. Also, the bark of balsam fir is smooth, and has many resin 'blisters', whereas spruce bark is rough and flakes off in scales. Balsam fir grows in richer swamps and uplands, from Alberta to Newfoundland, south to Iowa and Virginia. • Some birds, such as crossbills and grosbeaks, feed on black spruce seeds. Hares, squirrels, chipmunks, porcupines and deer feed on the bark, seeds and needles. • The genus name *Picea* means 'pitch,' and refers to the sticky sap, which can be used as a glue substitute. This sap was also chewed as gum and used for repairing boats. The wood was traditionally used to make spear handles. Today it is of great importance to the pulpwood industry. • The needles were made into a tea rich in vitamin C. Perhaps this is the famed *annedda* that the Iroquois used to miraculously save Cartier and his men in the early 1500s when they became deathly sick with scurvy. The bark was made into a medicinal salt. • Black spruce is called *ninaunduk* in Ojibway.

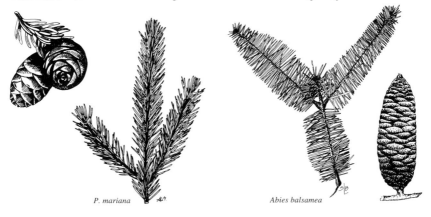

P. mariana   Abies balsamea

EASTERN WHITE CEDAR
THUYA OCCIDENTAL • *Thuja occidentalis*

GENERAL: Coniferous, evergreen tree, up to 20 m tall (much smaller in fens); branches layered, forming a cone-shaped crown; trunks twisted and contorted or straight and tapered, depending on habitat; twigs fan-shaped; **bark thin, shredding off in strips.**

LEAVES: Scale-like, in overlapping pairs, 2–4 mm long.

CONES: Unisexual, with both sexes on same tree; male cones small and deciduous; female cones oval, up to 1 cm long.

FRUITS: Double-winged seeds, released in late summer from 1-year-old female cones.

WHERE FOUND: In swamps, wet forests, calcium-rich uplands and fens and along shores; from Nova Scotia to southern Manitoba, south to Minnesota, Tennessee and North Carolina.

NOTES: White cedar seeds are a favourite food of pine siskins and common redpolls. Hares, beavers, porcupines and squirrels occasionally feed on the leaves. Deer browse extensively on the twigs and leaves. • The Ojibway called this tree *geezhig*, and they considered it one of the most useful trees in the forest. It provides a powerful, fragrant incense that is used to purify sacred objects, and it is used in ceremonies of the medicine lodge. White cedar was also used as a special offering to *Winabojo*. In legend, *Winabojo* talks of an ancient man who lives perpetually in the forest and wears a cedar tree on his head as an ornament, with its roots all around him. • The fragrant, resinous wood is popular for making fenceposts and rails, shakes and shingles, cedar strip canoes, closets and chests.

RED MAPLE
ÉRABLE ROUGE • *Acer rubrum*

GENERAL: Deciduous tree, up to 40 m tall, with watery, sweet sap; trunks up to 1 m in diameter (usually much smaller), with flaky or smoothish bark; twigs reddish.

LEAVES: Opposite, **broadly egg-shaped to nearly rounded, palmately 3–5-lobed**, squared or heart-shaped at base, pointed at tips, 7.5–15 cm long, irregularly toothed, bright green above, pale beneath; bright **crimson in autumn**.

FLOWERS: Small, red or sometimes yellowish, short-stalked, with 5 narrowly oblong petals; **in dense clusters** from twig sides; appearing much earlier than leaves, in March–April.

FRUITS: Pairs of 1-seeded, hairless, flattened, **winged keys** (samaras), slightly incurved, 2–3 cm long; wings 6–8 mm long, broadest above middle.

WHERE FOUND: In swamps and rich lowland areas; also on dry, rocky, south-facing hillsides; from southeastern Manitoba to Newfoundland, south to Texas and Florida.

NOTES: Silver maple (*A. saccharinum*) is similar to red maple, but its leaves are very deeply cut into 5 narrow lobes, its flowers lack petals and its young keys are woolly. These 2 species hybridize frequently, which further complicates identification. Silver maple is found in swamps and rich, moist bottomlands along streams and lakeshores, from Minnesota to New Brunswick, south to Oklahoma and Florida. • **Mountain maple** (*A. spicatum*) is a shrub or small tree. Its leaves have 3 (sometimes 5) relatively shallow, coarsely toothed lobes, and its greenish-yellow flowers appear after the leaves, in erect, dense, elongated clusters at the branch tips. Mountain maple grows in rich swamps, thickets and upland mixedwoods, where it often forms a conspicuous layer in the understorey. It is found from Saskatchewan to southern Labrador and Newfoundland, south to Iowa and northern Georgia. • Native peoples treated sore eyes with a lotion made from the pith of mountain maple twigs, and the bark was used to treat intestinal worms. • Pioneers boiled the bark of red maple and added copperas (iron sulphate) to the extracted tannin to made black ink. They made a cinnamon-coloured dye by adding alum instead of copperas, and they made black dye by adding both. Swedish settlers used both for dying linen. • Red maple sap was used to make syrup and sugar, though it is not as sweet nor as plentiful as that of **sugar maple** (*A. saccharum*).

• The light reddish-brown wood of red maple is hard, but not strong, so it is not considered an important timber species. It is used to make veneer, plywood, clothes hangers, clothes pins, kitchenware, crates and other items. Silver maple is widely planted as an ornamental. • Maples are preferred browse for moose, white-tailed deer, snowshoe hares, rabbits and beavers, especially in the winter. Ruffed grouse, songbirds, red squirrels, chipmunks and other small mammals eat the seeds, and ruffed grouse also eat the buds. It is commonly used by cavity-nesting birds and mammals, as it often has a hollow trunk. • The genus name *Acer* means 'sharp,' and it refers to the hardness of the wood, which the Romans used for spear shafts. Both the common and scientific names refer to the reddish colour of the buds, leafstalks, flowers, fruit and autumn leaves.

BLACK ASH
FRÊNE NOIR • *Fraxinus nigra*

GENERAL: Deciduous tree, up to 30 m tall, with **thick, opposite, ascending branches**; bark has corky ridges; small branches hairless, with **dark buds**; buds at tips **not separated** from lower side buds by a prominent gap.

LEAVES: Opposite, 25–40 cm long, **compound**, with 7–11 stalkless leaflets 10–12 cm long and 2–5 cm wide, often with **rusty hairs at base of leaflets**.

FLOWERS: Small and inconspicuous, in small dense clusters, unisexual, with male and female flowers on separate trees; appear in spring, before leaves.

FRUIT: Single, winged seeds (samaras), with flattened wing extending from near middle of seed, 2.5–4 cm long, 6–10 mm wide; mature in autumn and remain into winter.

WHERE FOUND: In low, rich forests and river valleys; from Newfoundland to southern Manitoba, south to North Carolina, Ohio and Delaware.

NOTES: Green ash (*F. pennsylvanica*) has stalked leaflets. It is found from Nova Scotia to southern Saskatchewan, south to Texas and Florida. • **White elm** (*Ulmus americana*) is another deciduous tree of rich swamps and stream floodplains, from Saskatchewan to Nova Scotia, south to Texas and Florida. It has simple, sharply double-toothed, 5–10 cm long, egg-shaped leaves that are abruptly sharp-pointed at the tip and unevenly matched at the base (1 side extends farther down the stalk than the other). Clusters (fascicles) of small flowers on slender, nodding stalks appear early in the spring, before the leaves. By May, these have produced oval, 1-seeded, flattened fruits (samaras) about 12 mm long, each with a distinctly veined, surrounding wing that is deeply notched at the tip. Unfortunately, Dutch elm disease has killed many of Ontario's elm trees over the past 50 years. • The Ojibway scraped the inner bark (cambium) of ash trees in long, fluffy layers and then cooked it. It was said to taste like eggs. • Black ash is used to make furniture and cabinets. The wet wood can be pounded with a wooden mallet to separate the growth rings, which can be used to weave baskets. • Many native peoples considered the wood of black ash a charm against serpents. This legend was passed on to early pioneers, who made cradles out of the wood to guard their babies against snakes. • Grosbeaks and mice eat black ash seeds, beavers and porcupines eat the bark, and deer and moose eat the twigs and leaves.

Ulmus americana

BALSAM POPLAR
PEUPLIER BAUMIER • *Populus balsamifera*

GENERAL: Deciduous tree, up to 25 m tall, **with thick, ascending branches; buds large, sticky, fragrant;** bark smooth and greenish when young, deeply furrowed with grey ridges when old.

LEAVES: Alternate, 7–13 cm long, egg-shaped, rounded at base, tapered to a pointed tip; edges with small rounded teeth.

FLOWERS: Tiny, in long, dense, **hanging, spike-like clusters (catkins),** unisexual, with male or female on separate trees; appear in spring, before leaves.

FRUITS: Egg-shaped pods, 5–8 mm long, containing many **tiny seeds** tipped **with tufts of long, silky hairs (pappus);** ripen in late spring or early summer, before leaves are fully expanded.

WHERE FOUND: In swamps and on shores; often associated with black spruce, balsam fir and trembling aspen; from Nova Scotia to Alaska, south to Oregon, Nebraska and Pennsylvania.

NOTES: Black willow (*Salix nigra*) is a common tree of riverbanks and other moist sites from North Dakota to New Brunswick, south to Texas and Florida. It can reach 30 m in height and its slender, tapered, lance-shaped leaves are 6–12 cm long and about 1–2 cm wide, with finely toothed edges. • Balsam poplar is distinguished from other species of *Populus* by its large sticky buds, egg-shaped leaves and preference for wet conditions. • Poplars offer a supermarket of food for wildlife. Birds, such as grouse and finches, and beavers eat the buds and catkins. Hares, muskrats, porcupines, squirrels and mice eat the bark, leaves and buds, and deer and moose browse on the twigs and leaves. • The wood has limited value for use in making plywood or pulpwood. The buds were once exported to Europe as 'tacamahaca,' where they were used as incense. When heated, the fragrant oils become airborne. Balsam poplar buds were also used medicinally in tinctures and salves. • The Ojibway call balsam poplar *zaudee*. They ate the gum under the bark, and they cooked the buds in lard to make a salve for treating burns, cuts, wounds and bruises or for use as a nasal decongestant.

P. balsamifera

Salix nigra

P. balsamifera

19

SAGE-LEAVED WILLOW
SAULE TOMENTEUX • *Salix candida*

GENERAL: Deciduous shrub, 1–1.5 m tall; young branches yellowish brown and hairy; older branches hairless and red.

LEAVES: Alternate, 3–10 cm long, **whitish green**, lance-shaped; veins impressed above, hidden by **dense white, woolly hairs below**; midrib often yellow; edges **rolled under**; stalks 3–10 mm long.

FLOWERS: Tiny, in dense, spike-like clusters (catkins), unisexual, with male and female catkins on separate plants; male catkins 1–2.5 cm long; female catkins 2–5 cm long; appear at same time as leaves.

FRUITS: Capsules, 5–8 mm long, hairy, stalkless, containing many, tiny seeds tipped with tufts of silky white hairs, in dense, spike-like clusters (catkins).

WHERE FOUND: In fens and wet meadows, on marshy lakeshores and around shallow pools behind beach dunes of the Great Lakes; from Newfoundland to Alaska, south to southern British Columbia, South Dakota and New Jersey.

NOTES: Two other common willows have toothless leaves and stalkless capsules, but lack woolly hairs on their leaves. • **Flat-leaved willow** (*S. planifolia*, also called *S. phylicifolia*) grows up to 3 m tall and has dark, glossy green, hairless leaves. Its catkins develop before its leaves, and the capsules have short, silky hairs. It is found from Alaska to Labrador, south to California and New Hampshire. • **Lowland pussy willow** (*S. discolor*, p. 23) grows up to 5 m tall and has variable leaves with leaf-like bracts (stipules). Its catkins develop before its leaves and the capsules, stalks and bracts are hairy. • Branches of willow catkins (pussy willows) are commonly collected in the spring, before the catkins have elongated, and used for decoration. • The name *candida* means 'white,' and it refers to the thick, white hair on the underside of the leaves.

BOG WILLOW
SAULE PÉDICELLÉ • *Salix pedicellaris*

GENERAL: Low, less than 1 m tall, deciduous, reddish shrub; low branches can root into moss, creating large low patches.

LEAVES: Alternate, oval to oblong, **2–5 cm long, leathery**, dark green above and whitened beneath, with a net-like pattern of veins on both sides; **midrib prominent, yellow to reddish brown**; stalks 2–6 cm long.

FLOWERS: Tiny, in dense, spike-like clusters (catkins), unisexual, with male and female catkins on separate plants; male catkins 0.5–2 cm long; female catkins same or slightly longer; appear at same time as leaves.

FRUITS: Hairless capsules above slightly hairy bracts, containing many tiny seeds tipped with tufts of silky white hairs, in catkins; stalks 2–4 mm long, slender.

WHERE FOUND: An excellent fen indicator; from Newfoundland to the Yukon Territory, south to Oregon, Iowa, Pennsylvania and New Jersey.

NOTES: The name *pedicellaris* refers to the distinctive long, slender stalks (pedicels) of the capsules.

SLENDER WILLOW
SAULE À LONG PÉTIOLE • *Salix petiolaris*

GENERAL: Deciduous shrub, 1.5–3.5 m tall; branches slender, erect to ascending.

LEAVES: Narrowly lance-shaped, tapered at both ends, 8–17 mm wide, finely and **evenly toothed**, slightly silky when young but soon hairless, green above, **whitish below** with a waxy bloom.

FLOWERS: Tiny, in **ovoid-cylindrical catkins**, unisexual, with male and female catkins on separate plants; male catkins about 2–3 cm long; female catkins 1–4 cm long, dense at first but becoming looser as capsule stalks elongate; appear in May, **before the leaves**.

FRUITS: Hairy capsules, 4–6 mm long, ovoid at base, blunt-beaked at tip; **stalks slender,** about half as long as capsule; **scales pale** brown or yellowish.

WHERE FOUND: In marshes, swamps, wet meadows, ditches and along shores of beaver ponds, lakes and rivers; often in dense thickets, particularly in lime-rich areas; also in dry upland forests; from British Columbia to Nova Scotia, south to Nebraska and New Jersey.

NOTES: Bebb's willow (*S. bebbiana*), also known as **beaked willow**, is very similar to slender willow, but its leaves are broader (egg-shaped and wider towards the tip), they have smooth or scalloped edges, and they usually have a prominent network of raised veins on the undersurface (especially when young). It grows in moist to wet habitats, including fens, swamps, lakeshores, riverbanks, alluvial flats and upland forests, from Alaska to Newfoundland, south to Arizona, Nebraska and Maryland. • **Sandbar willow** (*S. exigua*, also known as *S. interior*) has very long (3–15 cm), narrow (4–15 mm wide) leaves that are about 10 times as long as wide. They have widely spaced, regular teeth, and they are nearly stalkless. Sandbar willow spreads extensively, forming dense clumps on sandy lakeshores, floodplains, alluvial flats and riverbanks, and also in shallow water around ponds and at the edges of swamps. It is found from Alaska to James Bay and New Brunswick, south to California, Texas and Delaware. • Native peoples used the bark and roots of slender willow in a remedy to control bleeding, and they used the bark in a fever and headache remedy. They used the bark of sandbar and other willows to make fishnets and rope. • Moose, deer, beavers, snowshoe hares and muskrats browse on willows.

S. petiolaris

S. bebbiana

S. exigua

BALSAM WILLOW
SAULE BAUMIER • *Salix pyrifolia*

GENERAL: Deciduous shrub, 2–3 m tall, with a **distinctive balsam-like odour**.

LEAVES: Alternate, often purplish and translucent when young, firm, leathery and dark green when mature, **pear- to lance-shaped, heart-shaped at base**, with a distinctive **net-like vein pattern** and short, regular teeth; stalks 0.5–2 cm long.

FLOWERS: Tiny, in dense, spike-like clusters (catkins), unisexual, with male and female catkins on separate plants; male catkins 2–5 cm long; female catkins 2–5 cm long or slightly longer; appear with leaves.

FRUITS: Hairless capsules, 5–8 mm long, above hair-like bracts, containing many, tiny seeds tipped with tufts of silky white hairs, in catkins.

WHERE FOUND: In swamps and ditches and on sandy shorelines; from Newfoundland to British Columbia, south to Minnesota, New York and New England.

NOTES: Several other wetland willows have regularly toothed leaves but lack the balsam-like odour of balsam willow. • **Shining willow** (*S. lucida*) has green leaves that are shiny on both surfaces and long-pointed at their tips. Its capsules are hairless and its catkins appear with the leaves. Shining willow is common in swamps, marshes, wet fields and ditches and on lakeshores, from Newfoundland to Saskatchewan, south to South Dakota, Ohio and Maryland. • **Autumn willow** (*S. serissima*) has dark green leaves that are glossy above, whitened beneath and have gland-tipped teeth. Its capsules are hairless and its catkins appear in midsummer, well after the leaves have come out. Autumn willow is also common, usually in marshes and fens, and is found from Newfoundland to the Northwest Territories, south to Colorado, South Dakota and New Jersey. • **Bebb's willow** (*S. bebbiana*), also called **beaked willow**, is 1–6 m tall. It has variable leaves and hairy capsules, and its catkins appear at the same time as the leaves. It is common and widespread in thicket swamps and on shores, as well as in dry, upland forests, from Alaska to Newfoundland, south to Arizona, Nebraska and Maryland. • **Peach-leaved willow** (*S. amygdaloides*) has finely toothed leaves with long, slender points, very similar to those of shining willow, but lacking the small glands found on the leaf stalks. The branches of this tall shrub or small tree tend to arch and nod at the tips. Peach-leaved willow grows on lakeshores and riverbanks and at the edges of swamps and marshes, from Alaska to Newfoundland, south to California, New Mexico and Maryland. It flowers in April or May, before the leaves have expanded. • The species name *pyrifolia* refers to its 'pear-shaped leaves.' An earlier name, *S. balsamifera*, referred to the balsam-like odour of the leaves.

S. pyrifolia   *S. lucida*

LOWLAND PUSSY WILLOW
SAULE DISCOLORE • *Salix discolor*

GENERAL: Deciduous shrub or small tree, up **to 7 m tall** with trunks up to 30 cm in diameter; branches essentially hairless.

LEAVES: Elliptic to lance-shaped, tapered at both ends, 7–12 cm long, 1.5–4 cm wide, **regularly round-toothed** to nearly toothless, **hairless when mature** but sometimes hairy when young, bright **green above, whitened below** with a waxy bloom, slender-stalked.

FLOWERS: Tiny, in dense, **thick, cylindrical catkins,** unisexual, with male and female catkins on separate plants; female catkins 2.5–7 cm long; appear in March–May, **well before leaves**.

FRUITS: Woolly, narrowly cone-shaped capsules, 5–6 mm long, much longer than stalks; **scales brownish purple,** with long, glossy hairs.

WHERE FOUND: Damp meadows, swamps, thickets and wet ditches and along lakeshores and streambanks; from British Columbia to Newfoundland and Labrador, and south to Idaho, Missouri and Maryland.

NOTES: **Missouri willow** (*S. eriocephala,* also called *S. rigida*) is closely related to lowland pussy willow, and it is sometimes considered a variety of that species. It is generally hairier than lowland pussy willow, its catkins are denser and more silvery-silky, its leaves have reddish hairs on the lower side of the veins (even at maturity), and its small branches are more or less hairy. It grows on river flats, the banks of streams and ponds, wet roadsides and in thickets and swamps, from Newfoundland to the Yukon, south to California, with disjunct populations in Alaska and Florida. • **Flat-leaved willow** (*S. planifolia,* also called *S. phylicifolia*) grows up to 3 m tall and has hairless, toothless leaves that are glossy green above and whitish with a waxy bloom beneath when mature. Its catkins develop before its leaves and the capsules have dense, silky hairs. Flat-leaved willow grows in a wide range of moist habitats, including fens, lakeshores, streambanks, rocky shores, sedge marshes, swamps and treed fens, from Alaska to Labrador, south to California and New Hampshire. • Willow bark contains salicylic acid, the active ingredient in Aspirin, and it was formerly used as a treatment for rheumatism and other ailments. • In England, pussy willow twigs with catkins are called 'palms,' and they are collected by children to be used for church decoration on Palm Sunday. • Many willow species are browsed by moose. • The species name *discolor* means 'of 2 colours,' and it refers to the leaves.

SWEET GALE
MYRIQUE BAUMIER • *Myrica gale*

GENERAL: Deciduous shrub, up to 1.5 m tall, **fragrant**; stems hairy, gland-dotted.

LEAVES: Alternate, 3–6 cm long, toothed at tip, dotted on upper and lower surfaces with yellow glands.

FLOWERS: Small, in dense, spike-like clusters (catkins), unisexual, with **male and female flowers on separate plants**; appear before or with leaves.

FRUITS: Gland-dotted achenes above 2 corky bracts, in ovoid, cone-like, female catkins 10–12 mm long.

WHERE FOUND: On shores and in fens; from Newfoundland to Alaska, south to Oregon, Minnesota and North Carolina.

NOTES: Grouse, catbirds, chickadees, crows and bluebirds eat sweet gale seeds, and deer browse on the twigs and leaves. Sweet gale often forms dense, low thickets on rocky shores, where it provides cover for many species. • Sweet gale is a nitrogen-fixer. Its root nodules contain nitrogen-fixing bacteria that live in a symbiotic relationship with the shrub. • Sweet gale seeds are dispersed by water, floating on their corky bracts. • Native people used the leaves and seeds to make tea and to season meat. The boiled seeds produce a yellow dye. Sweet gale has been used as an insect repellent and as a hop substitute for flavouring malt liquors. • The genus name *Myrica* means 'fragrant.' The fragrance of the leaves and stems is especially noticeable when they are crushed.

DWARF BIRCH, SWAMP BIRCH
BOULEAU NAIN • *Betula pumila*

GENERAL: Densely branched, deciduous shrub, up to 2 m tall; branches covered with small, crystalline specks (glands).

LEAVES: Alternate, numerous, **dark green** (turning scarlet to brown in autumn), 2–4 cm long, almost **round, coarsely toothed and veined**, often with many **yellow dots (glands) on both sides.**

FLOWERS: Tiny, **in dense, spike-like clusters (catkins)**, unisexual, with male and female catkins on same shrub, develop in late summer, pollinated next spring.

FRUITS: Flat, 2–4 mm long nutlets, in papery scales of 1–2 cm long, cone-like, female catkins; shed together with scales in spring.

WHERE FOUND: In fens and open conifer swamps and on lakeshores; a good fen indicator, never found in bogs; from Newfoundland to Alaska, south to Oregon, Iowa and New Jersey.

NOTES: Dwarf birch is eaten by deer and moose. • The Ojibway burned the tiny cones as an aromatic to treat swollen mucous membranes of the nose and throat. Women drank a tea made from the cones to give them strength during childbirth. The branches are used as support ribs in baskets made of sweet grass (*Hierochloe odorata*). A flammable substance called bitumen can be recovered through distillation of the bark of this and other birches.

SPECKLED ALDER
AULNE RUGUEUX • *Alnus incana*

GENERAL: Tall, deciduous shrub, up to 4 m tall; **twigs covered with white specks** (lenticels); **pith 3-sided; winter buds stalked**; in tight clumps.

LEAVES: Alternate, oval, 6–10 cm long, 3–6 cm wide, **doubly toothed**.

FLOWERS: Tiny, in dense, spike-like clusters (catkins), unisexual, with male and female catkins on same plant; male catkins long, hanging, scaly; female catkins small, oval, erect; develop in late summer, pollinated next spring.

FRUITS: Winged nutlets, in rounded, 1 cm long, persistent, **woody cones that are stalkless or very short-stalked**.

WHERE FOUND: On lakeshores and streambanks and in swamps; from Newfoundland to Alaska, south to California, Iowa and Pennsylvania.

NOTES: Speckled alder was formerly known as *A. rugosa*. • **Green alder** (*A. viridis*, also known as *A. crispa*) can be distinguished by its sticky branches and young leaves, its stalkless buds, its long-stalked woody cones and its finely toothed leaves. It is typically found in fresh or dry upland forests, but it also grows in wet disturbed areas and along rocky lakeshores, from Newfoundland to Alaska, south to California and South Carolina. • Alder seeds are an important food for redpolls, goldfinches and pine siskins. Grouse feed on the buds and seeds. Beavers, rabbits and, more rarely, deer and moose eat the twigs and leaves. The dense growth of the stems offers important cover for wildlife. • Alders have nitrogen-fixing bacteria in nodules on the roots. • Native people have used the inner bark of alders for making yellow or red dyes.

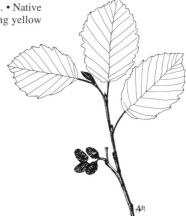

A. viridis

MOUNTAIN HOLLY
NÉMOPANTHE MUCRONÉ • *Nemopanthus mucronatus*

GENERAL: Deciduous shrub, up to 3 m tall; branches either long and purple with few leaves, or short and grayer with many leaves.

LEAVES: Alternate, bright green, oval, with an **abrupt point,** toothless or with sharp teeth at tip; **stalks purple, slender.**

FLOWERS: Unisexual, 2 mm wide, **on thread-like stalks from leaf axils**; mid-spring.

FRUITS: Purple-red berries, 6 mm wide, 4–5-seeded, on **thin, thread-like stalks**; late summer.

WHERE FOUND: In swamps and moist forests and on bog borders; from Newfoundland to northwestern Ontario, south to Illinois and Virginia.

NOTES: Another member of the holly family, **holly winterberry** (*Ilex verticillata*), also called **black alder,** is found in swamps, peatlands and wet thickets, from Ontario to Newfoundland, south to Minnesota and Georgia. These 2 genera are separated on the basis of their flowers, which have oblong to egg-shaped, slightly joined petals in winterberry and narrow, separate petals in mountain holly. Also, winterberry can easily be distinguished by its larger (5–7.5 cm vs. 1.2–5 cm long), sharply toothed (rather than toothless), leathery (rather than thin) leaves. The orange-red, berry-like fruits of winterberry are stalkless and smooth, whereas those of mountain holly are stalked and slightly ribbed. • A European species, English holly (*I. aquifolium*), has evergreen leaves with sharp spines on their edges. It is used for Christmas decorations. • Wildlife use of mountain holly is unknown, but waterfowl, songbirds and many mammals extensively eat the fruits of holly winterberry. • Native people used the buds and bark of mountain holly as a remedy for fever and inflammation, and they called it 'fever bush.' The fruit is said to be bitter-tasting and to act as a laxative. • The name *Nemopanthus* is of Greek origin, meaning 'thread-foot flower,' and it refers to the long, thread-like flower and fruit stalks.

Ilex verticillata

NARROW-LEAVED MEADOWSWEET
CHICOUTÉ • *Spiraea alba*

GENERAL: Deciduous shrub, 1.5 m tall; branches numerous, dull, yellowish brown, often with longitudinal ridges and papery, peeling bark.

LEAVES: Alternate, crowded, oval to lance-shaped, **at least 3 times as long as wide**, 3–6 cm long, 1–2 cm wide, with many fine, **sharp teeth**; stalks 2–6 mm long.

FLOWERS: White, 5–8 mm wide, 5-parted, numerous, in short-hairy, **narrow pyramids at branch tips**; throughout summer.

FRUITS: Small, smooth, few-seeded capsules in clusters of 5–8; often remain on plant through winter.

WHERE FOUND: In wet meadows, swamps and ditches and along streambanks and lakeshores; from Newfoundland to Alberta, south to North Dakota, Ohio and Delaware.

NOTES: Broad-leaved meadowsweet (*S. latifolia*) is less common than narrow-leaved meadowsweet, and it is usually found on dry sand and rock, along lakeshores and riverbanks, from Manitoba to Newfoundland, south to the northeastern United States. It is distinguished by its broader leaves (less than 3 times as long as wide), its essentially hairless flower cluster, and its red-brown to purple-brown twigs. • **Steeple-bush** (*S. tomentosa*), also called **hardhack**, is easily recognized by the dense, whitish, woolly hairs on the underside of its leaves and on its twigs. Unlike narrow-leaved and broad-leaved meadowsweet, steeple-bush has pink or purple (rarely white) flowers. It grows on shores, marshes, swamps, fields and roadsides, from Manitoba to Nova Scotia, south to Mississippi and North Carolina. • Narrow-leaved meadowsweet is sometimes eaten by white-tailed deer. • This shrub has long been used to make a flavourful tea. Native people used it as a tonic and as a treatment for nausea. • The species name *alba* means 'white,' and it refers to the flowers.

S. tomentosa

BLACK CHOKEBERRY
ARONIE À FRUIT NOIR • *Aronia melanocarpa*

GENERAL: Deciduous shrub, 1–3 m tall, with slender branches.

LEAVES: Alternate, toothed, egg-shaped, with a tooth-like tip; veins curve toward tip (similar to dogwood leaf veins); **hair-like glands along midvein.**

FLOWERS: White, 5-parted, 5–10 mm in diameter, in small clusters (cymes) at branch tips; early summer.

FRUITS: Black berries, 6–10 mm in diameter; mid- to late summer.

WHERE FOUND: On shrubby lakeshores, in swamps and rarely in upland pine forests; from Newfoundland to northwestern Ontario, south to Texas and Florida.

NOTES: Grouse, chickadees, cedar waxwings, foxes, rabbits, squirrels, mice, deer and black bears eat the berries. • Native people used the berries to make a tea to treat the common cold. • The species name *melanocarpa* means 'black berries.'

ALDER-LEAVED BUCKTHORN
NERPRUN À FEUILLES D'AULNE • *Rhamnus alnifolia*

GENERAL: Deciduous shrub, usually less than 1 m tall; branches grey, with **small ridges when mature**; forms low thickets.

LEAVES: Alternate, up to 10 cm long and 5 cm wide, oval, with small rounded teeth; **veins strong**, sunken, giving leaf a corregated appearance.

FLOWERS: 3 mm wide, with **5 yellow-green sepals** and no petals, 1–3 in lower leaf axils; late spring or early summer.

FRUITS: Black, short-stalked berries, 6–8 mm wide, with 1–3 seeds; late summer.

WHERE FOUND: In conifer swamps and fens and on shores; from Newfoundland to southern British Columbia, south to California, Ohio and New Jersey.

NOTES: These fruits are eaten in small quantities by birds, black bears and hares. In late winter, they may be the only berries available, and there are often a few hungry, reluctant birds sitting among them. The leaves are lightly eaten by deer. • Alder-leaved buckthorn is an alternate host for the fungus that causes oat rust. • Buckthorns contain a strong laxative, which can be **fatal** if eaten in large quantities. • The genus name *Rhamnus* means 'prickly shrub,' referring to a closely related European species, *R. cathartica*, which has spines. The species name *alnifolia* refers to the alder-like leaves.

SMALL CRANBERRY
AIRELLE CANNEBERGE • *Vaccinium oxycoccos*

GENERAL: Low, trailing, evergreen shrub with wiry stems, usually **forming thick, bushy clumps or mats**.

LEAVES: Alternate, oval, 3–8 mm long, shiny green above, whitened beneath; edges curved under; stalkless or short-stalked.

FLOWERS: Pale-pink, 4-parted, with petals curved back, shooting star-like, nodding on slender stalks, solitary from leaf axils or in clusters of 2–6 at stem tips; early to mid-summer.

FRUITS: Many-seeded, dry, sour, red berries with tough skins, 6–8 mm in diameter; late summer to autumn, and remain on plant through winter.

WHERE FOUND: In bogs, fens and conifer swamps, often on peat moss hummocks; from Newfoundland to Alaska, south to Oregon, Michigan and North Carolina.

NOTES: **Large cranberry** (*V. macrocarpon*, also known as *Oxycoccus macrocarpus*) is very similar to small cranberry, but its leaves are larger (6–17 mm long), and they have blunt or rounded (rather than pointed) tips. Also, the berries are oblong (rather than round) and much larger (10–20 mm in diameter), with 2 small, leaf-like bracts above middle of the stalk (in small cranberry these bracts are at or below the middle). Large cranberry usually grows in open peatlands, but it can also be found in swamps and on wet shores. Its range extends from Minnesota to Newfoundland, south to Arkansas and North Carolina. • **Mountain cranberry** (*V. vitis-idaea*) has evergreen leaves that are strongly whitened and black-dotted beneath. Its berries are red and sour. Mountain cranberry is found in treed bogs and fens, in conifer swamps and on rocky coasts from Newfoundland to Alaska, south to Lake Superior and New England. • Cranberries are said to be of little food value to wildlife, although grouse have been observed eating them. • When cooked with a sweetener, small cranberry fruits make an excellent sauce for turkey, chicken or game meat.

Wild or cultivated large cranberries are the main source of commercial cranberries in eastern North America. • The Native people called cranberries *toca*. Cranberries were prized for their ability to keep for a long time, and they were collected in large quantities. They were used to treat dysentery, wounds, inflamed tumours and nausea. Cranberry juice is said to effectively combat urinary tract infections, and it is used and prescribed today for that purpose. • The name *oxycoccos* is from the Greek *oxys*, 'sharp,' and *kokkos*, 'berry,' referring to the flavour of the fruit.

V. macrocarpon

BOG BILBERRY
AIRELLE DES VASES • *Vaccinium uliginosum*

GENERAL: Low, less than 60 cm tall, branching, deciduous shrub; young branches reddish brown; older branches greyish brown to purplish.

LEAVES: Alternate, elliptic to rounded, leathery, **5–25 mm long, 5–15 mm wide**; veins on underside form a **distinctive net-like pattern**; edges slightly rolled under.

FLOWERS: White or pink, 4–5-parted, urn-shaped, 5–6 mm long, solitary or **in nodding clusters at branch tips**; mid-summer.

FRUITS: Blue to black berries, 4–6 mm wide, many-seeded; late summer.

WHERE FOUND: In bogs and poor fens and on lakeshores; from Newfoundland to Alaska, south to Lake Superior and New York.

NOTES: Bog bilberry is a northern species that also grows in the cold Lake Superior microclimate. It is distinguished from our other common blueberries (**low sweet blueberry [*V. angustifolium*, p. 31]** and **velvet-leaved blueberry [*V. myrtil-loides*, p. 31])** by its smaller leaves. • Native peoples of the far north eat bog bilberries fresh or dried. All northern species of blue berries are edible and provide vitamins A, B and C, as well as traces of calcium, phosphorus and iron. Flavours vary from fruity to insipid and from tart to sweet, depending on the species and the time of year. • Northern native peoples recommend blueberry tea as a refreshing drink and a remedy for diarrhea. • Many wild birds and mammals feed on *Vaccinium* species. The berries are an important part of the diet of black bears, chipmunks, tanagers, and grouse, and the leaves and twigs are browsed by deer, elk and hares.

SHRUBS

VELVET-LEAVED BLUEBERRY
AIRELLE FAUSSE-MYRTILLE • *Vaccinium myrtilloides*

GENERAL: Deciduous shrub, 30–60 cm tall, with warty stems and soft-hairy twigs; usually **in thick, bushy clumps or mats.**

LEAVES: Softly hairy, alternate, oval, 2.5–5 cm long, **toothless**; stalks short and hairy.

FLOWERS: 5-parted, 4–5 mm wide, white to pink or purple, in clusters at stem tips; early to mid-summer.

FRUITS: Rounded berries, 4–7 mm wide, **blue, usually with a waxy film;** mid- to late summer.

WHERE FOUND: In acidic habitats, such as poor fens, swamps and dry uplands; from Newfoundland to British Columbia, south to Montana and Virginia.

NOTES: Low sweet blueberry (*V. angustifolium*) is distinguished by the fine teeth on its leaf edges and by its hairless stems and leaves. It grows in gravelly or sandy wooded areas from Newfoundland to Manitoba, south to Iowa and West Virginia. • Huckleberry (*Gaylussacia baccata*) is closely related to the blueberries, but its fruit is a black, berry-like drupe, containing 10 1-seeded nutlets (rather than a many-seeded, blue berry). The young flowers and leaves of huckleberry are thickly covered with shiny, resinous globules, which give the plant a clammy feeling. Huckleberries are sweet, though seedy. They grow in peatlands and moist, sandy to rocky woods from central Ontario to Nova Scotia, south to Louisiana and Georgia. • Blueberries are very important to wildlife during the summer. Grouse, starlings, robins, sparrows, thrushes and orioles, feed heavily on blueberries. Bears obviously love blueberries (you only have to see bear dung once during late summer to verify this), but other mammals, such as chipmunks, squirrels, mice and skunks, also find these sweet, juicy berries very palatable. Deer, rabbits and hares browse on the leaves, branches and berries throughout the year. • Many people pick and eat blueberries in northern Ontario. Native people traditionally collected and dried them in large quantities for use in the winter months. In Ojibway, blueberries are called *meenun*, and the roots were said to have great medicinal value, especially as a tea taken before childbirth to facilitate a safe delivery. • The name *myrtilloides* comes from the resemblance of this species to the European blueberry (*V. myrtillus*).

Gaylussacia baccata

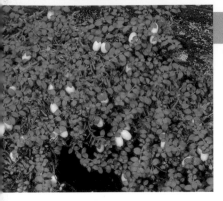

CREEPING SNOWBERRY
CHIOGÉNE HISPIDE • *Gaultheria hispidula*

GENERAL: Creeping, evergreen shrub; stems covered with bristly, brown hairs.

LEAVES: Alternate, less than 1 cm long, oval, green above, with **brown bristles beneath**.

FLOWERS: White, 2–3 mm long, bell-shaped, **4-lobed**, in leaf axils; spring to early summer.

FRUITS: White, juicy berries with **evergreen flavour**; mid- to late summer.

WHERE FOUND: In treed bogs, swamps and moist conifer forests on moss, rocks and logs; from Newfoundland to British Columbia, south to Idaho, Michigan and North Carolina.

NOTES: Creeping snowberry could be confused with **mountain cranberry** (*Vaccinium vitis-idaea*), which has white hairs on its stems, larger leaves and red berries. It is also similar to **small cranberry** (*V. oxycoccos*, p. 29), which has leaves that are hairless beneath and red berries. • Grouse, black bears, mice and deer feed on these berries. • Creeping snowberry berries have a lemony, evergreen flavour, and traditionally they have been used to make teas, liqueurs, jams and jellies. The leaves have a minty, wintergreen flavour. The berries and the leaves should be eaten sparingly, because they contain methyl salicylate, which **can be toxic** when taken in large amounts. This plant has been used as a stimulant and for the prevention of cavities and treatment of asthma.

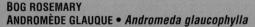

BOG ROSEMARY
ANDROMÈDE GLAUQUE • *Andromeda glaucophylla*

GENERAL: Low, **evergreen** shrub, usually less than 60 cm tall; branches erect.

LEAVES: Alternate, leathery, stiff, linear, **2–5 cm long, with edges rolled under**, tipped with a small spine, strongly **whitened and hairy beneath**.

FLOWERS: Pink, 5–6 mm wide, urn-shaped, in **nodding clusters at stem tips**; early summer.

FRUIT: Rounded **capsules**, less than 6 mm across, with a **persistent style** in depression at tip; mid- to late summer.

WHERE FOUND: In bogs and fens; from Newfoundland to Manitoba, south to Minnesota and New Jersey.

NOTES: Dwarf bog rosemary (*A. polifolia*) is a circumpolar, arctic and alpine species whose range extends south to southern British Columbia and to the coasts of Hudson Bay and James Bay in Ontario. It is a smaller shrub, with leaves that are whitened but not hairy beneath. • The Ojibway used both the fresh and dried leaves for making tea. However, the leaves contain the **poison** andromedotoxin, which is dangerous to livestock. • The species name *glaucophylla* means 'whitened leaves.'

LEATHERLEAF
CASSANDRE CALICULÉ
Chamaedaphne calyculata

GENERAL: Low, **evergreen** shrub, less than 1 m tall; often in dense clumps.

LEAVES: Alternate, 1.5–5 cm long, progressively smaller toward tip of branch, firm, leathery, oval, green above, with **white-brown scales beneath.**

FLOWERS: Nodding, 5-parted, with petals fused into a **white urn, in elongated, 1-sided clusters at branch tips**; late spring to early summer.

FRUITS: Small capsules, remain on branches for several years, contain many small seeds; mid- to late summer.

WHERE FOUND: In bogs, fens, and conifer swamps and on lakeshores; from Newfoundland to Alaska, south to Iowa and Georgia.

NOTES: Leatherleaf seeds are an important winter staple for grouse. Deer and moose also eat them. • Native people boiled the leaves to make a tea to treat fevers. This beverage could be **dangerous**. Any boiling would extract andromedotoxin from the leaves, making the drink **poisonous**. • The seed capsule forms a kind of rattlebox, and wind, rain or animals shake the seeds out through slits that open along the seams. • The genus name *Chamaedaphne* means 'creeping on the ground,' but this phrase is misleading because leatherleaf is usually erect, and it can grow up to 1 m tall.

LABRADOR TEA
THÉ DU LABRADOR • *Ledum groenlandicum*

GENERAL: Low, **evergreen shrub**, less than 1 m tall; **young stems with woolly, orange-brown hairs**.

LEAVES: Alternate, leathery, thick, 2–5 cm long, dark green, with dense, rusty-coloured hairs beneath and down-rolled edges, fragrant when crushed.

FLOWERS: White, 1 cm wide, 5-parted, in small clusters at branch tips; early summer.

FRUITS: Oval **capsules**, 5–6 mm long, **tipped with slender, 5–7 mm long styles**, contain many seeds, split open from bottom to tip; empty capsules remain on plant for several years; mid- to late summer.

WHERE FOUND: Most abundant in conifer swamps, also in treed bogs and fens and in moist depressions in upland forests; from Newfoundland to Alaska, south to Oregon, Minnesota and New Jersey.

NOTES: Labrador tea is adapted to survive in harsh environments. The leaves are evergreen, allowing the plant to minimize the loss of precious nutrients in nutrient-poor bogs. At times, water supply can be very limited in bogs. The leaves are exposed to hot, dry conditions by late spring, but the roots (which take up water for the plant) can remain frozen well into the summer. The leathery texture, down-curved edges and hairy surfaces of the leaves are adaptations to minimize water loss. • Deer apparently feed on the leaves of Labrador tea. • Labrador tea is a well-known beverage, and it is quite thirst-quenching. To the Ojibway, *neebeeshaubo* (tea) was of great importance, and it was the preferred beverage they drank while travelling. The Cree called Labrador tea *karkarpukwa*. • These plants are said to have medicinal properties, and they have been used to treat headaches and restlessness. An insect repellent made from crushed Labrador tea leaves, alcohol and glycerine is apparently very effective for keeping mosquitoes away and relieving insect bites.

BOG LAUREL
KALMIA À FEUILLES D'ANDROMÈDE
Kalmia polifolia

GENERAL: Low, **evergreen** shrub, less than 1 m tall, with flattened, 2-edged branches; **buds lack scales**.

LEAVES: Opposite, oval or elliptic, 1–5 cm long, **stalkless**, evergreen, **leathery**; edges rolled under.

FLOWERS: Pink, saucer-shaped, showy, 9–15 mm wide, 5-parted, in loose, clusters at stem tips; early to mid-summer.

FRUITS: Rounded capsules, up to 6 mm long, **with a persistent style** and many small seeds; mid- to late summer.

WHERE FOUND: In poor fens, bogs and conifer swamps; from Newfoundland to Alaska, south to Oregon, Minnesota and New Jersey.

NOTES: Sheep laurel (*K. angustifolia*) has flowers in the axils of year-old leaves, rounded stems and leaves that are stalked, whorled and broadly oval. It grows around lakes and in bogs, from Newfoundland to northeastern Ontario, south to Michigan, Virginia and Georgia. • The Cree called bog laurel *wesukapup*, and they used it to make a tea and a tonic. • There have been several reported **human poisonings** (not fatal) in which similar symptoms of vertigo, loss of sight, coldness of extremities, nausea and vomiting were experienced after ingesting this plant. This reaction is due to a **glycoside**, arbutoside, and an **andromedotoxin** found in the plant. **Do not eat this plant.** • Laurels are said to be of little value to wildlife. Perhaps this is due to their toxicity, for they have reportedly killed cattle, sheep, goats and horses. • The genus name *Kalmia* commemorates Pehr Kalm, one of the first botanists to collect and record the flora of Canada.

K. angustifolia

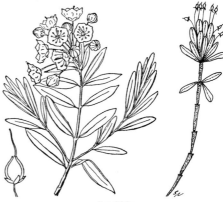

K. polifolia

RED-OSIER DOGWOOD
CORNOUILLER STOLONIFÈRE • *Cornus stolonifera*

GENERAL: Deciduous shrub, 2–3 m tall; **stems bright red** (usually recognizable at a distance), with **large, white pith**; in dense thickets, often spreading by stolons.

LEAVES: Opposite, 5–10 cm long, oval; veins curve toward leaf tip.

FLOWERS: White, 6–8 mm wide, in **flat-topped clusters** at branch tips, 5-parted; early summer.

FRUITS: White or bluish berries, 6 mm in diameter; late summer.

WHERE FOUND: In marshes, swamps, wet fields and ditches and on shores; from Newfoundland to Alaska, south to California, Nebraska, Ohio and New England.

NOTES: Round-leaved dogwood (*C. rugosa*) has green warty bark usually streaked with purple and larger (7–15 cm long), broadly oval to round leaves. It is found from Nova Scotia to southern Manitoba, south to Iowa, Ohio and Virginia. • **Silky dogwood** (*C. amomum*), with its purplish twigs and opposite, strongly veined leaves, resembles red-osier dogwood. However, its leaves are silky-downy on their lower surface, and its berries are light blue, whereas the leaves of red-osier dogwood are hairless or only sparsely hairy, and its berries are usually white. Older branches have a slender, brownish pith rather than a large, whitish pith as in red-osier dogwood. Silky dogwood grows on wet sites, including low-lying woods, marshes, ditches, thickets and streambanks, from North Dakota to New Brunswick, south to Oklahoma and Georgia. • **Gray dogwood** (*C. racemosa*) also resembles red-osier dogwood, but it has greyish (rather than red to purple) branches with slender, brownish pith, and its leaves have 3–4 (rather than 5–7) pairs of veins. It has white berries, usually on bright red stalks (rather than green or purplish as in silky dogwood). Gray dogwood grows in wet thickets and along streambanks from southeastern Manitoba to Quebec and Maine, south to Oklahoma, Kentucky and Maryland. • Red-osier dogwood is an important shrub to many wildlife species, particularly as a winter staple. Wood ducks, grosbeaks and thrushes eat the berries, which account for as much as 50% of the evening grosbeak's diet. Rabbits, skunks, squirrels, chipmunks, mice, deer, moose and bears eat the fruit, wood and leaves. • The Ojibway used the inner bark to make *kinnikinnik*, an ancient, mystical tobacco, used before real tobacco became readily available. This was smoked in pipes at many ceremonies, used as a peace offering, and sprinkled on grave boxes to help the dead journey to the spiritland. The stems were also used to make baskets and red dye.

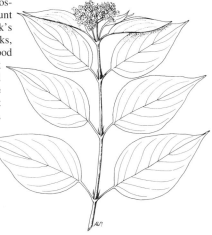

BRACTED HONEYSUCKLE
CHÈVREFEUILLE INVOLUCRÉ • *Lonicera involucrata*

GENERAL: Deciduous shrub, 2–3 m tall; mature branches squarish, with a **solid, white pith and peeling bark.**

LEAVES: Opposite, oval, with pointed tips, 5–15 cm long, hairless above, slightly downy beneath; stalks 1 cm long.

FLOWERS: Showy, fragrant, yellow, 10–15 mm long, **funnel-shaped or tubular and lobed,** in opposite **pairs from leaf axils;** stalks over 1 cm long, above 2 large, pointed bracts; early summer.

FRUIT: Paired, few-seeded, **purple-black berries above** 4 broad, green to dark purple, **leaf-like bracts; stalks over 1 cm long**; mid- to late summer.

WHERE FOUND: In swamps and damp woods and on shores; from Quebec to Alaska, south to California, Wisconsin and Michigan.

NOTES: Canada fly honeysuckle (*L. canadensis*) grows in swamps and upland forests, from Nova Scotia to northern Ontario, south to Iowa, Ohio and North Carolina. Its distinctive features include slender (not leaf-like) bracts below pairs of red berries, long (over 1 cm) stalks on flowers and fruits, and blunt tips, hairless surfaces and hairy edges and stalks on the leaves. • Honeysuckle flowers are pollinated by humming-birds and long-tongued insects, such as hawkmoths. Occasionally, a bee chews a hole at the base of a flower to obtain nectar. Birds eat the fruits, and then disperse the undigested seeds. • The name *involucrata* refers to the distinctive bracts (involucre) below the fruits and flowers.

MOUNTAIN FLY HONEYSUCKLE
CHÈVREFEUILLE VELU • *Lonicera villosa*

GENERAL: Deciduous shrub, less than 1 m tall; **branches have solid, white pith and reddish-brown bark that peels off** in layers.

LEAVES: Opposite, elliptic, rounded at tip, 2.5–6 cm long, hairy on both surfaces, with fringe of hairs along edges; **stalks less than 3 mm long.**

FLOWERS: Showy, fragrant, yellow, **tubular or funnel-shaped,** lobed, in **opposite pairs from leaf axils; stalks less than 1 cm long;** early summer.

FRUITS: Few-seeded, blue berries; mid- to late summer.

WHERE FOUND: In conifer swamps and intermediate to rich fens and on lakeshores; never in bogs; from Newfoundland to Manitoba, south to Minnesota and New England.

NOTES: Swamp fly honeysuckle (*L. oblongifolia*) has similar, but less-hairy leaves, flowers and fruits on stalks over 1 cm long, pairs of red to purple berries and gray-brown bark. It grows from Nova Scotia to Saskatchewan, south to Minnesota and Pennsylvania. • Native people have used honeysuckle berries to make tea and juice and some people use them to make jams and jellies. • The name *villosa* refers to the distinctively hairy leaves.

L. oblongifolia

HIGHBUSH CRANBERRY
VIRONE TRILOBÉE • *Viburnum trilobum*

GENERAL: Erect, **deciduous shrub**, 1–4 m tall; branches smooth.

LEAVES: Opposite, deeply 3-lobed (**maple-leaf-like**), pointed at tips, rounded at base, sometimes wider than long, coarsely toothed; stalks with 2 glands at tip.

FLOWERS: White, of 2 types, in 7.5–10 cm broad, **flat-topped clusters** (cymes); **inner flowers** small, relatively **inconspicuous**, fertile; **outer flowers showy**, sterile, with flat, 11–25 mm wide corollas; June–July.

FRUITS: Red, **translucent**, berry-like drupes with a single, flattened seed, very acidic, 8–10 mm in diameter.

WHERE FOUND: Wet, rich, clay or silt soil in rich swamps and thickets and on floodplains, streambanks and shores; from British Columbia to Newfoundland, south to Washington, Illinois and Pennsylvania.

NOTES: This species was previously known as *V. opulus*. • Two other species of *Viburnum*, **withe rod** (*V. cassinoides*) and **nannyberry** (*V. lentago*) are also common on rich wet sites in Ontario, but their leaves are simple and egg-shaped rather than 3-lobed, and their flower clusters lack the showy, radiant, outer flowers of highbush cranberry. The 6–9 mm long 'berries' of withe rod are borne in spreading clusters at the tips of long stalks, whereas those of nannyberry are 10–15 mm long, and grow in stalkless, branched clusters. Also, the leaves of withe rod have irregular, rounded teeth, and those of nannyberry have sharp, regular teeth. Withe rod grows on moist or acid soils of lakeshores, riverbanks, thickets, swamps and peatland edges, from Ontario to Newfoundland, south to Wisconsin and Alabama. Nannyberry grows in swamps and thickets and on shorelines and floodplains, from southeastern Saskatchewan to southwestern Quebec, south to Colorado, Missouri and Georgia. • **Mooseberry** (*V. edule*) has 3-lobed leaves, but they are not as deeply cleft, and they have finely toothed edges. Mooseberry has only a few, small fertile flowers borne on short, 2-leaved, side branches. It grows in damp woods and along lakeshores and streambanks, as well as in upland forests, from Alaska to Newfoundland, south to Oregon, Colorado and Pennsylvania. • Highbush cranberry could be confused with **mountain maple** (*Acer spicatum*, p. 17) but the leaves of that maple have longer stalks, and they are sharply toothed. • Highbush cranberry was called 'cramp bark' by settlers who made a tea from the bark to treat stomach and menstrual cramps, spasms and mumps. Native people chewed mooseberry twigs and swallowed the juice to relieve sore throats.

V. cassinoides

V. edule

SWAMP RED CURRANT
GADELLIER AMER • *Ribes triste*

GENERAL: Deciduous, low shrub less than 1 m tall; **stems lacking spines and bristles.**

LEAVES: Alternate, 4–10 cm long, coarsely toothed, **palmate**, with 3–5 lobes, often **clustered on short branches**; stalks 2.5–6 cm long.

FLOWERS: Greenish purple, up to 6 mm wide, in small nodding clusters on long stalks from leaf axils; early summer.

FRUITS: Many-seeded, smooth, bright red berries, 6–9 mm wide; mid- to late summer.

WHERE FOUND: In swamps and moist forests and on river- and streambanks; from Newfoundland to Alaska, south to Oregon, South Dakota and Virginia.

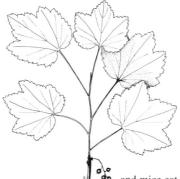

NOTES: There are 2 other wetland species of *Ribes* with stems lacking prickles (unarmed) in this region. **Skunk currant (*R. glandulosum*)** smells like a skunk, its leaves lack resin dots, and it has glandular, red berries. It grows from Newfoundland to Alaska, south to British Columbia, Minnesota, Ohio and North Carolina. **Northern wild black currant (*R. hudsonianum*)** has resin-dotted leaves and black berries. It is found in moist sites from western Quebec to Alaska, south to California, Michigan, and southern Ontario • Gulls, catbirds, grouse, squirrels, chipmunks and mice eat swamp red currants. Deer and moose eat the leaves and twigs. • The seeds are dispersed by birds and mammals, and the seed coat is softened as it passes thorough the digestive tract, which hastens germination. The Ojibway call currants and gooseberries *zhaubominuk*, which means 'the seed that goes through.' • Swamp red currant berries were traditionally used for making jams and jellies, and they are still prized for their good taste and high pectin content (pectin is needed to make preserves gel). They are called *meeshidjiminuk* in Ojibway and were used as a winter staple or preserve. The Ojibway used swamp red currant to cure such things as urinary disorders and eye sores. • The genus name *Ribes* is taken from the Arabic for 'acid-tasting,' and it refers to the tartness of the berries.

R. hudsonianum

R. glandulosum

R. glandulosum

BRISTLY BLACK CURRANT
GADELLIER LACUSTRE • *Ribes lacustre*

GENERAL: Deciduous, low shrub, less than 1.5 m tall; **stems bristly, often with pairs of larger thorns at branch nodes**.

LEAVES: Alternate, 4–8 cm long, **palmately veined**, deeply 3–5-lobed, with **scattered glandular hairs on stalk and undersurface**, coarsely and bluntly toothed.

FLOWERS: Up to 6 mm wide, saucer-shaped, yellow-green to pink, in nodding clusters on long stalks from leaf axils; early summer.

FRUITS: Purple-black berries, with **stalked glands**, 9–12 mm wide, foul-tasting, many seeded; mid- to late summer.

WHERE FOUND: In moist forests and swamps, on rocky lakeshores and on river- and streambanks; from Newfoundland to Alaska, south to California, Minnesota, Tennessee and New York.

NOTES: Two similar *Ribes* species of northern Ontario wetlands are **wild gooseberry** (*R. hirtellum*) and **bristly wild gooseberry** (*R. oxyacanthoides*). Both have spines and prickles that are shed with the bark to reveal a smooth stem. The spines at the branch nodes are 1 cm long, and the berries do not have glands or bristles. These 2 species can be distinguished by their leaves. Wild gooseberry leaves lack glands, whereas bristly wild gooseberry leaves have glands on the veins on the underside. Bristly wild gooseberry flowers have a spicy fragrance. Wild gooseberry is found from Newfoundland to Saskatchewan, south to Montana, Minnesota and Pennsylvania, whereas bristly wild gooseberry grows from Hudson Bay to Alaska, south to Montana, Minnesota, Ohio and Pennsylvania • Identification of *Ribes* species can be difficult when hybridization occurs. This is especially true of hybrids of wild gooseberry and bristly wild gooseberry. • *Ribes* species are the alternate host for white pine blister rust (*Cronartium ribicola*), the fungus that causes blister rust in white pines. • Gulls, catbirds, grouse, squirrels, chipmunks, mice and other animals eat bristly black currant berries. Deer and moose eat the leaves and twigs. • The species name *lacustre* means 'of lakes,' and it refers to the wetland habitat.

R. hirtellum

R. lacustre

CLOUDBERRY, BAKE-APPLE
CHICOUTÉ • *Rubus chamaemorus*

GENERAL: Deciduous, erect shrub, 10–30 cm tall; **branches lack prickles (unarmed).**

LEAVES: Alternate, **palmate, with 5–7 rounded lobes**, finely toothed; stalks 2–8 cm long.

FLOWERS: Showy, white, solitary, **2–3 cm wide, long-stalked, 5–7-parted**; early to mid-summer.

FRUITS: 1–2 cm wide, pale red to yellow, raspberry-like clusters of pulpy or fleshy berries (drupelets); mid- to late summer.

WHERE FOUND: In bogs, fens and black spruce swamps; usually on peatmoss; from Newfoundland to Alaska, south to Lake Superior and Nova Scotia.

NOTES: Cloudberries are a popular fruit in northern Canada and Newfoundland, but these fruits are rarely found in the southern part of northern Ontario, even where cloudberry plants are plentiful. The soft fruits are best eaten fresh as they soon turn mushy if they are carried for any distance, but too many cloudberries on an empty stomach can cause cramps. In the past, native peoples preserved this fruit in oil or fat. Today, cloudberries are usually cooked with a bit of sugar, frozen, or canned. • Some northern peoples commonly used cloudberry leaves as a tobacco stretcher. Large quantities of leaves were boiled with a plug of tobacco and then dried. These leaves were said to taste just as good as tobacco, so 1 plug became many. Sometimes cloudberry leaves were mixed with kinnikinnick leaves or the inner bark of red osier dogwood for smoking. • The species name is derived from the Greek *chamai*, 'on the ground,' and the Latin *morum*, 'mulberry.'

WILD RED RASPBERRY
RONCE DU MONT IDA • *Rubus idaeus*

GENERAL: Deciduous shrub; stems (canes) up to 1.5 m tall, with **prickles** (armed), biennial; spreading by stolons.

LEAVES: Alternate, **compound, with 5–7 egg-shaped**, toothed, **3–6 cm long leaflets**.

FLOWERS: White, 1–1.5 cm wide, with 5 petals, **in clusters** of 2–5 at stem tips; June–July.

FRUITS: Tiny, red, fleshy berries (druplets) in rounded, 1 cm long clusters (raspberries), easily removed from centre (receptacle) when ripe; July–August.

WHERE FOUND: In open disturbed areas, ditches, hardwood swamps and shrubby thickets and on shores; from Newfoundland to Alaska, south to California, Tennessee and North Carolina.

NOTES: Also known as *R. strigosus*. • Many birds, including grosbeaks, robins, sparrows, thrushes, orioles, catbirds and grouse, eat these berries. Black bears, chipmunks, squirrels, raccoons, mice, hares and rabbits also eat wild red raspberries. Rabbits and hares take cover in and feed extensively on, raspberry stems during the winter. Deer and moose heavily browse on the leaves and stems throughout the year. Raspberries are important as food, but they also provide dense, prickly cover and protection for small animals and nesting habitat and materials for birds. • The Ojibway called raspberries *miskominuk*. Native peoples had many uses for this plant. They collected and dried the berries, and used them as a winter staple, for making pemmican and for preparing jams and preserves. They dried the leaves, and used them to make tea. They used the roots and other plant parts to treat dysentery, eye sores, stomach aches and bad coughs, and to regulate labour pains and prevent miscarriages. Raspberry bark is high in tannins, and it is a good astringent. • While exploring Canada in 1619, Champlain found abundant raspberries and blueberries. He remarked, 'We also found it was almost as if God had wished to bestow a gift of some sort on this barren unfriendly country.'

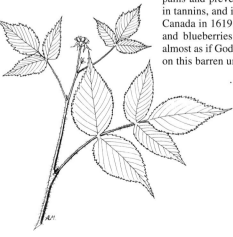

DWARF RASPBERRY
RONCE PUBESCENTE • *Rubus pubescens*

GENERAL: Deciduous, low or trailing shrub, 10–30 cm tall, with perennial **runner-like stems and erect, herbaceous, leafy branches**; erect, flowering branches softly hairy, **lacking prickles** (unarmed); trailing, sterile branches have whip-like ends, often rooting at joints (nodes).

LEAVES: Alternate, compound, with **3 or 5 leaflets**; leaflets diamond- or egg-shaped, 2–7 cm long, **tapering to a point**, sharply toothed, hairless.

FLOWERS: White or occasionally pale pink, with 5 petals (6–10 mm long), on slender, hairy stalks in loose clusters of 1–3 at branch tips; late spring to early summer.

FRUITS: Tiny, dark red, fleshy berries (druplets) in **rounded clusters (raspberries)**, not easily separated from centre (receptacle); mid- to late summer.

WHERE FOUND: In upland forests and swamps and along lakeshores, creeks and rivers; from Newfoundland to the Northwest Territories, south to Washington, South Dakota, Pennsylvania and New Jersey.

NOTES: **Northern dwarf raspberry** (*R. acaulis*, p. 45) is very similar, but it lacks trailing stems, and it has less sharply pointed leaflets and deeper pink flowers with petals that are 10–20 mm long. • **Swamp dewberry** (*R. hispidus*) is also a low, trailing 'raspberry' with 3 (rarely 5) leaflets per leaf. However, unlike dwarf raspberry, it has somewhat woody stems armed with dense, weak, backward-bent bristles. Also, its leaflets are more rounded than those of dwarf raspberry, and they are thicker and often somewhat shiny. The fruits of swamp dewberry are small, sour and almost black when ripe. This small, creeping shrub grows in swamps, wet meadows and woods from Wisconsin to Nova Scotia, south to Illinois and North Carolina. • The leaves of dwarf raspberry are easily confused with those of **strawberries** (*Fragaria* spp., p. 76), but strawberries have more rounded leaf tips, and they lack woody, trailing stems. • Dwarf raspberry fruits are edible, but they are smaller, more difficult to find and less desirable than those of **wild red raspberry** (*R. idaeus*, p. 43). • The species name *pubescens* refers to the pubescent or downy hairs covering the stems.

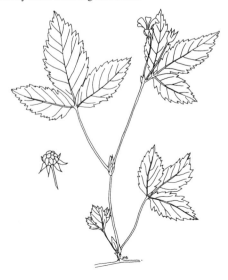

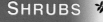

NORTHERN DWARF RASPBERRY, NAGOONBERRY
RONCE ACAULE • *Rubus acaulis*

GENERAL: Deciduous shrub, 5–10 cm tall, with erect leafy branches, lacking prickles (unarmed); from a tufted base.

LEAVES: Alternate, compound, **with 3 leaflets**; leaflets rounded or abruptly pointed at tip, 1–4.5 cm long, toothed; stalks longer than uppermost leaflets, slightly hairy; leaf-like bracts (stipules) oval, sometimes sheath-like.

FLOWERS: Solitary, light to dark pink, showy, 5–7-parted, 2 cm wide; stalks long and finely hairy; early to mid-summer.

FRUITS: Red, 1 cm wide, clusters of pulpy or fleshy berries (druplets); mid- to late summer.

WHERE FOUND: In fens, black spruce swamps and alder thickets and along creeks and rivers; from Newfoundland to Alaska, south to Lake Superior.

NOTES: Another low-growing *Rubus* species is **dwarf raspberry** (*R. pubescens*, p. 44). It is distinguished by its 10–30 cm tall flowering stems, white flowers, long, whip-like stems and long, sharp leaf points (not abruptly tapered). • Grosbeaks, robins, sparrows, thrushes, orioles, catbirds, grouse, black bears, chipmunks, squirrels, raccoons, mice and hares, feed on raspberries. • These tiny raspberries are edible but difficult to find. Perhaps the animals close to the forest floor find them first.

SHRUBBY CINQUEFOIL
POTENTILLE FRUTESCENTE • *Potentilla fruticosa*

GENERAL: Low, deciduous, bushy shrub, less than 1 m tall; young stems red and hairy; **older plants have peeling grey bark**.

LEAVES: Alternate, hairy, 1–2 cm long, **compound**, usually with 5 leaflets; **3 upper leaflets often united at base; stalks long and hairy, with 2 papery leaf-like bracts (stipules) at base.**

FLOWERS: Pale yellow, 5-parted, 2–3 cm wide, 1 to few at branch tips; throughout summer.

FRUITS: Clusters of achenes enclosed by persistent sepals; late summer to early autumn.

WHERE FOUND: Along riverbanks, lakeshores and rocky coastlines, on sand dunes and in fens; from Newfoundland to Alaska, south to California, South Dakota, Pennsylvania and New Jersey.

NOTES: Marsh cinquefoil (*P. palustris*, p. 77) has larger leaves and purple flowers, and it lacks woody stems. • Shrubby cinquefoil seeds are occasionally eaten by sharp-tailed grouse. Hares and rabbits eat the leaves. • The medicinal qualities of cinquefoils are known by herbalists throughout North America, Europe and Asia. These plants have been used to make a mouthwash, and to treat weak bowels, bleeding, sore throats and fevers.

PRICKLY WILD ROSE
ROSIER ACICULAIRE • *Rosa acicularis*

GENERAL: Low, prickly, deciduous shrub, up to 1.5 m tall; branchlets reddish, densely covered with slender prickles.

LEAVES: Alternate, compound, with 5–7 oval, toothed leaflets, each 2–5 cm long; **2 leaf-like bracts (stipules)** at base of stalk.

FLOWERS: 3–7 cm wide, pink, fragrant, with **5 petals and 5 sepals**; early to mid-summer.

FRUITS: Bright red, hairless, fleshy hips, 1–2 cm thick, with many achenes; mid- to late summer.

WHERE FOUND: On shores and in wet meadows, swamps and upland open woods; from Quebec to Alaska, south to Idaho, Minnesota and Vermont.

NOTES: Swamp rose (*R. palustris*) grows in swamps, meadows, marshes and fens and on shores, from Minnesota to Nova Scotia, south to Arkansas and Florida. It is easily distinguished from prickly wild rose by the glandular stalks of its flowers and fruits, and by its widely spread to backward-bent sepals, which fall off at maturity. Also, the broad-based, curved prickles of swamp rose occur only at the nodes of the twigs, and there are no bristles between the nodes. • Animals eat rose hips and buds throughout the winter. Birds, such as grouse and vireos, eat the hips and buds. Deer, black bears, beavers, rabbits, hares, skunks, squirrels and mice eat the fruits, stems and leaves. • Rose hips are very rich in vitamin C, and they can be made into tea or jam. However, the stiff hairs on the achenes inside the fruit are **very irritating** to the bowels. Native people used the hips as a remedy for indigestion and heartburn. • The species name *acicularis* means 'needle-shaped,' and it refers to the slender prickles on the stems.

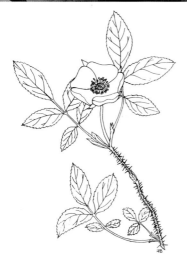

COMMON JUNIPER, GROUND JUNIPER
GENÉVRIER COMMUN • *Juniperus communis*

GENERAL: Low, coniferous, evergreen shrub, less than 1.5 m tall; in broad patches, often dead at the centre.

LEAVES: Stiff, prickly needles 6–18 mm long, with **white stripe on upper surface** (often hidden by upcurved edges), **in 3s.**

CONES: Unisexual, with male and female cones on separate plants, in axils of 1-year-old leaves.

FRUITS: Fleshy, **berry-like cones, waxy, bluish white**, 6–13 mm wide, containing 1–3 seeds.

WHERE FOUND: On sandy or rocky lakeshores and in open fens, old meadows and sometimes open woods; from Newfoundland to Alaska, south to California and Georgia.

NOTES: Creeping juniper (*J. horizontalis*) is found in rich fens and along sandy or rocky lakeshores, from Newfoundland to Alaska, south to Wyoming and New York. It can be distinguished by its scale-like leaves and trailing branches. • When birds eat the berry-like cones, their digestive tract breaks down the seed coat, which hastens the seed's germination after it is distributed in the bird's droppings. Common juniper 'berries' are a favourite food of many songbirds, such as grosbeaks and finches. Up to half of a cedar waxwing's diet can be juniper 'berries.' Small mammals, such as chipmunks, squirrels and mice, occasionally eat the fruits and deer browse the twigs and leaves. • Juniper is called *kauwaunzh* by the Ojibway and *wakinakim* by the Cree. The bark was boiled and crushed to make an antiseptic dressing for wounds. • The fruits are crushed in alcohol to flavour gin.

J. communis

J. horizontalis

J. horizontalis

47

HERBS

This section includes non-woody flowering plants, except the aquatics and the grass-like plants. It is important to remember that many floating and submerged herbs are included in the 'aquatics' section of this guide (pp. 153–89), and the plant you are trying to find may be described there. There are over 1900 different herbs in Ontario. This guide includes common species that grow in Ontario wetlands.

A chart based on the following leaf and flower features is included to aid in the identification of unknown plants.

Leaf Features

opposite    *whorled*    *alternate*    *basal*

*compound
3 leaflets*    *compound
more than 3 leaflets*    *toothed*    *not toothed*

Flower Features

tubular    *dense heads
(aster-like)*    *tiny & crowded* 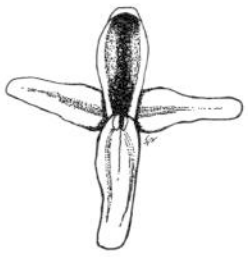   *petals not alike*

*petals alike
3 parts*    *petals alike
4 parts*    *petals alike
5 parts*    *petals alike
6–7 parts*

Only the species described in this book are included in the chart. Use the chart as follows:
1. Note leaf features.
2. Note flower features. Even if there are no flowers and you cannot determine colour, there may still be sepals or petals on the fruits. If the petals or sepals are similar (each part identical) they can be counted together. Heads of tiny, petal-like flowers (e.g. thistles and asters) that appear to have more than six parts go into the 'aster-like' category. Flowers that have parts that are not identical (e.g. orchids and peas) fit into the 'petals-not-alike' category.
3. Look for the right combination of characteristics on the chart. This should narrow the choice to a few species.

One plant may be found in several different categories in the chart because some characteristics, such as flower colour and leaf shape, can vary.

FLOWERS	LEAVES						
	Opposite/Whorled		Alternate		Basal	Compound	
	toothed	not toothed	toothed or dissected	not toothed		3 leaflets	more than 3 leaflets
White Flowers							
tubular			*Campanula* (p.97)	*Campanula* (p.97)			
dense heads or aster-like	*Eupatorium* (p.100)		*Aster* (pp.98–9) *Cirsium* (p.100)	*Aster* (pp.98–9)	*Petasites* (p.101)		*Cirsium* (p.100)
tiny & crowded	*Mentha* (p.95) *Lycopus* (p.94)	*Galium* (pp.96–7)		*Maianthemum* (p.57)	*Tofieldia* (p.58)		*Sium* (p.87) *Cicuta* (pp.86–7) *Thalictrum* (p.71)
petals not alike	*Chelone* (p.94)		*Viola* (pp.80–3)	*Platanthera* (pp.62–3)	*Platanthera* (pp.62–3) *Viola* (pp.80–3) *Spiranthes* (p.64) *Malaxis* (p.63)		
petals alike 3 parts		*Galium* (pp.96–7)			*Tofieldia* (p.58)	*Trillium* (p.58)	
petals alike 4 parts		*Galium* (pp.96–7) *Cornus* (p.88)		*Maianthemum* (p.57)			*Cardamine* (p.68) *Nasturtium* (p.68)
petals alike 5 parts	*Epilobium* (p.85)	*Epilobium* (p.85)	*Anemone* (p.70) *Campanula* (p.97)	*Campanula* (p.97) *Geocaulon* (p.67) *Parnassia* (p.75) *Tiarella* (p.74)	*Mitella* (p.74) *Orthilia* (p.89) *Pyrola* (p.89) *Drosera* (p.73) *Parnassia* (p.75)	*Fragaria* (p.76) *Menyanthes* (p.91) *Coptis* (p.71)	*Anemone* (p.70)
petals alike 6–7 parts		*Trientalis* (p.89)		*Maianthemum* (p.57)	*Tofieldia* (p.58)	*Coptis* (p.71)	
Violet/Purple or Blue Flowers							
tubular			*Campanula* (p.97)	*Campanula* (p.97)			
dense heads or aster-like	*Eupatorium* (p.100)		*Aster* (pp.98–9) *Cirsium* (p.100)	*Aster* (pp.98–9)			
tiny & crowded	*Eupatorium* (p.100) *Mentha* (p.95)						*Thalictrum* (p.71)
petals not alike	*Scutellaria* (p.95) *Mimulus* (p.93)	*Listera* (p.65)	*Viola* (pp.80–3)		*Viola* (pp.80–3)	*Arisaema* (p.56)	
petals alike 3 parts					*Iris* (p.59)		
petals alike 4 parts		*Rhexia* (p.85)					
petals alike 5 parts	*Mentha* (p.95)	*Asclepias* (p.92) *Lythrum* (p.84) *Triadenum* (p.79) *Agalinis* (p.93)	*Campanula* (p.97)	*Campanula* (p.97)	*Sarracenia* (p.72)	*Menyanthes* (p.91)	*Potentilla* (p.77) *Geum* (p.77)
petals alike 6–7 parts		*Lythrum* (p.84)			*Iris* (p.59)		
Orange Flowers							
petals not alike			*Impatiens* (p.80)				

FLOWERS	LEAVES						
	Opposite/Whorled		Alternate		Basal	Compound	
	toothed	not toothed	toothed or dissected	not toothed		3 leaflets	more than 3 leaflets
Yellow Flowers							
dense heads or aster-like	*Bidens* (p.103)	*Bidens* (p.103)	*Solidago* (p.102)	*Solidago* (p.102) *Euthamia* (p.102)			*Megalodonta* (p.103)
petals not alike			*Impatiens* (p.80)	*Platanthera* (pp.62–3) *Cypripedium* (p.60)	*Platanthera* (pp.62–3)		
petals alike 3 parts				*Scheuchzeria* (p.55)	*Xyris* (p.55) *Scheuchzeria* (p.55)		
petals alike 5 parts		*Lysimachia* (5–6 petals) (p.90) *Hypericum* (p.78)	*Caltha* (5–9 petals) (p.69)	*Caltha* (p.69)			
Pink or Red Flowers							
dense heads or aster-like	*Eupatorium* (p.100)		*Cirsium* (p.100)	*Aster* (pp.98–9)	*Petasites* (p.101)		
tiny & crowded	*Asclepias* (p.92) *Mentha* (p.95)			*Polygonum* (p.67)			
petals not alike	*Chelone* (p.94) *Listera* (p.65)			*Platanthera* (pp.62–3) *Pogonia* (p.62)	*Arethusa* (p.61) *Cypripedium* (p.60) *Calopogon* (p.61)		
petals alike 3 parts						*Trillium* (p.58)	
petals alike 4 parts		*Decodon* (p.84) *Rhexia* (p.85)					
petals alike 5 parts	*Epilobium* (p.85)	*Epilobium* (p.85) *Agalinis* (p.93) *Decodon* (p.84) *Triadenum* (p.79) *Asclepias* (p.92)			*Pyrola* (p.89) *Sarracenia* (p.72) *Drosera* (p.73)		*Potentilla* (p.77)
Green or Brown Flowers							
tiny & crowded		*Galium* (pp.96–7)	*Boehmeria* (p.66) *Laportea* (p.66) *Urtica* (p.66)				
petals not alike		*Listera* (p.65)		*Platanthera* (pp.62–3)	*Platanthera* (pp.62–3)	*Arisaema* (p.56)	
petals alike 3 parts		*Galium* (pp.96–7)		*Scheuchzeria* (p.55)	*Scheuchzeria* (p.55)		
petals alike 4 parts		*Galium* (pp.96–7)					
petals alike 5 parts				*Geocaulon* (p.67)	*Orthilia* (p.89) *Mitella* (p.74)		
petals alike 6 parts					*Triglochin* (p.54)		

Wildflower Photo Guide

Not all the wildflowers described in this guide are included on these pages, but there should be enough representatives of the major flower types to guide you to the appropriate section of the book. Some shrubs and aquatics are also included in this photo guide.

p. 59   p. 81   p. 95   p. 98   p. 95

p. 84   p. 101   p. 92   p. 188   p. 72

p. 77   p. 65   p. 89   p. 74   p. 87

p. 64   p. 62   p. 94   p. 97   p. 91

p. 57   p. 100   p. 98   p. 160   p. 76

p. 70

p. 71

p. 75

p. 89

p. 88

p. 83

p. 58

p. 60

p. 62

p. 61

p. 93

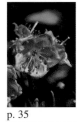
p. 35

p. 46

p. 93

p. 85

p. 81

p. 61

p. 100

p. 90

p. 102

p. 55

p. 90

p. 181

p. 90

p. 45

p. 69

p. 184

p. 103

p. 60

p. 80

GREATER ARROWGRASS
TROSCART MARITIME • *Triglochin maritimum*

GENERAL: Grass-like, perennial herb, 20–80 cm tall; from creeping rhizomes.

LEAVES: Basal, **up to 30 cm long and only 1–2 mm wide,** flaring to a **wide sheath** (resembling the fletching on an arrow).

FLOWERS: Stalked, 2–5 mm wide, with 6 greenish petals/sepals, each with a large, stalkless anther, in a **10–40 cm long, narrow cluster (raceme).**

FRUITS: Oval capsules, 5 mm long and **2.5 mm thick,** usually **remaining on the plant for a long time**.

WHERE FOUND: Fairly common fen indicator; less common along streams or on gravelly lakeshores; from Newfoundland to Alaska, south to California, Texas, Ohio and Pennsylvania.

NOTES: Greater arrowgrass is a conspicuous plant, as it towers above sedges in rich fens. • **Slender arrowgrass (*T. palustre*)** is distinguished by its slender growth and long, thin capsules (6 times longer than wide). It is found from Newfoundland to Alaska, south to California, Nebraska and Pennsylvania. • Another plant that might be confused with greater arrowgrass is **podgrass (*Scheuchzeria palustris*, p. 55),** which is also found in fens, but which produces widely divergent flowers and leaves all the way up its stems. Also, its leaves have a pore at the tip. • Greater arrowgrass seeds are eaten by waterfowl, such as black ducks and mallards. • The species name *maritimum* means 'growing by the sea' and refers to this plant's other habitat, which is in salt marshes along the Atlantic coast.

T. maritimum

T. palustre

T. palustre

PODGRASS, NUTGRASS
SCHEUCHZÉRIE PALUSTRE • *Scheuchzeria palustris*

GENERAL: Yellow-green, perennial herb, 10–40 cm tall; 'zig-zag' stems; from a creeping rhizome.

LEAVES: Alternate, somewhat tubular, with a **distinctive pore at tip,** erect, 5–30 cm long, 1–3 mm wide, with dilated, open sheaths; clustered near base and separated on stem.

FLOWERS: Yellow-green, with 3 petals and 3 sepals about 3 mm long and 6 stamens; in an elongated cluster (raceme) above a bract at stem tip; early to mid-summer.

FRUITS: Pod-like capsules (follicles) in 3s, 4–8 mm long, containing 1 or 2 seeds; July–August.

WHERE FOUND: In open bogs and poor fens; from Newfoundland to Alaska, south to California, Nebraska and New Jersey.

NOTES: Despite the common name, this is not a grass. It is the only member of the podgrass family in North America, but it is related to the **arrowgrasses (*Triglochin* spp.,** p. 54). • This genus was named for Swiss botanist John Jakob Scheuchzer. The species name *palustris* means 'of the marsh or swamp.'

MOUNTAIN YELLOW-EYED GRASS
XYRIS DES MONTAGNES • *Xyris montana*

GENERAL: Perennial herb; **stems densely tufted, seldom reaching 30 cm** in height; from somewhat branching rhizomes.

LEAVES: Dark green, **4–15 cm long, only 1–2 mm wide,** with edges folded over and fused at tip.

FLOWERS: Yellow, with **3 broad, clawed petals** and 3 sepals, stalkless, **on small (1.5–3 mm long), head-like spikes** of overlapping, **scale-like bracts**; mid- to late summer.

FRUITS: Oblong capsules, splitting into 3 parts, underneath scale-like bracts; seeds 1 mm long, ribbed.

WHERE FOUND: In rich, open fens, usually in large peatland complexes; occasionally on sandy or muddy lakeshores; from Newfoundland to Lake Superior, south to Pennsylvania and New Jersey.

NOTES: Southern yellow-eyed grass (*X. difformis*) is a larger (10–50 cm tall), stouter species, with much broader leaves (up to 15 mm wide at the middle) and flattened, conspicuously winged stems. It has smaller seeds (0.5–0.6 mm, rather than 0.8–1 mm long), and the tips of the side sepals are strongly toothed, rather than toothless or nearly so. It is found on sandy or gravelly shores and in peatlands and wet meadows from Ontario to Quebec, south to Texas and Florida. • Yellow-eyed grass is not a grass but belongs to the family Xyridaceae. It is a species of the Atlantic Coastal Plain, locally distributed in the Great Lakes Region. • The genus name *Xyris* was derived from a Greek term meaning 'razor,' referring to the 2-edged appearance of the leaves at their tips.

55

JACK-IN-THE-PULPIT
ARISÉMA TRIPHYLLE • *Arisaema triphyllum*

GENERAL: Perennial **herb,** with **intensely acrid** juice; flowering stems single, 25–80 cm tall; from a **wrinkled, tuberous corm**.

LEAVES: Erect, 1–2, slender-stalked, compound, **with 3 elliptic-egg-shaped,** 7.5–15 cm long **leaflets**.

FLOWERS: Tiny, unisexual, with male and female on separate plants, **crowded in a dense cluster at base of a yellowish, cylindrical to club-shaped,** 5–7.5 cm long **structure** (spadix), surrounded and **hidden by a large, green- and purple-striped bract** (spathe) that curls around to form a **funnel with a broad, pointed flap arching forward over the spadix.**

FRUITS: Small, **smooth, red berries**, in conspicuous, **rounded clusters** 2.5–7.5 cm long; June–July.

WHERE FOUND: In rich hardwood or cedar swamps, moist thickets, rich woods and floodplains; from southeastern Manitoba to Nova Scotia, south to Kansas and South Carolina.

NOTES: This species is also known as *A. atrorubens*. • Immature plants without flowers may be confused with **poison ivy (*Rhus radicans*)**, but the 3 leaflets in poison ivy are stalked and their veins are not joined along the edge. • Jack-in-the-pulpit plants can alternate sexes from one year to the next, depending on the amount of food (starch) accumulated in the corm the previous year. A large corm storing abundant food is required to produce a female flower and fruit. A male flower is produced if the corm is small, as pollen production requires less energy. If the corm is very small, only a leaf and no flowers are produced. • Jack-in-the-pulpit plants contain irritating **crystals of calcium oxalate**, which are **intensely acrid**. These sharp-pointed crystals embed themselves in mucous membranes, where they are both a chemical and mechanical irritant. • Calcium oxalate crystals are destroyed by long exposure to heat and/or dry air, and native peoples buried the plants and built a longlasting fire over them. The dried, starchy roots were then ground into flour and used to make bread. European settlers use Jack-in-the-pulpit in syrup or ointment to treat colic, flatulence, asthma, whooping cough, sores and ringworm. The dried root was said to be an excellent remedy for hoarseness and loss of voice, but the Hopi claimed 1 teaspoon of powdered, dried root to half a glass of water would induce temporary sterility—2 teaspoons would cause permanent sterility. • The leaves are rarely eaten by insects or mammals, but wood thrushes have been known to eat Jack-in-the-pulpit berries. • The name of the arum family comes from the Arabian word *ar*, meaning 'fire,' and it refers to the severe pain and burning sensation caused by calcium oxalate crystals. Jack-in-the-pulpit is known as *zhaushaugomin* (spongy berry) in Ojibway. The common name comes from the leaf-like spathe which is folded over as a hood to form the 'pulpit', sheltering the tube-like preacher (the spadix), popularly known as the 'Jack.' The genus name *Arisaema* comes from the Greek *haima* or blood, in reference to the spotted leaves of some species. The species name *triphyllum* refers to the 3 leaflets.

CANADA MAYFLOWER, WILD LILY-OF-THE-VALLEY
MAÏANTHÈME DU CANADA • *Maianthemum canadense*

GENERAL: Low, perennial herb, 5–20 cm tall; often forming large patches from slender, branching, spreading rhizomes.

LEAVES: 1–3, dark green, 3–10 cm long, alternate, **heart-shaped at base**, with many parallel veins.

FLOWERS: Star-shaped, 4–6 mm wide with **4 white tepals**, in a **crowded cluster above leaves**; spring to early summer.

FRUITS: Pale red, speckled berries, 3–4 mm in diameter, with 1–2 seeds; mid-summer.

WHERE FOUND: In a wide range of habitats including dry to moist coniferous or deciduous forests, swamps and sand beaches; from Newfoundland to British Columbia, south to Iowa, Tennessee and Georgia.

NOTES: Three-leaved Solomon's seal (*M. trifolium*, below) is distinguished from Canada mayflower by its 6-parted flowers and narrower, lance-shaped leaves. • Grouse, chipmunks and mice eat the berries, and rabbits and hares eat the leaves. • The berries have a bitter taste and are cathartic. Native people traditionally used Canada mayflower to make a tea to treat sore throats and headaches. • The genus name *Maianthemum* means 'May flower.'

THREE-LEAVED SOLOMON'S SEAL
SMILACINE TRIFOLÉE • *Maianthemum trifolium*

GENERAL: Slender, erect, 10–30 cm tall perennial herb; from creeping rhizomes.

LEAVES: 3, lance-shaped, 6–12 cm long, 1–4 cm wide, **stalkless, ascending,** with parallel veins.

FLOWERS: White, less than 1 cm wide, 6-parted, **in slender loose clusters above leaves**; early summer.

FRUITS: Dark red, speckled berries, 1–2-seeded; mid-summer.

WHERE FOUND: In fens, swamps and bogs; from Newfoundland to the Yukon Territory, south to southern Alberta, Minnesota and New Jersey.

NOTES: This species was formerly known as *Smilacina trifolia*. • Three-leaved Solomon's seal may be confused with **Canada mayflower** (*M. canadense*, above), which has heart-shaped leaves and 4-parted flowers. • Grouse, thrushes and mice eat the berries, and hares, deer and moose eat the leaves. • These berries can cause intestinal problems if eaten in quantity. Traditionally, they were used by native people to treat headaches, back pains and coughs. • The species name *trifolium* means 'with 3 leaves.'

STICKY FALSE ASPHODEL
TOFIELDIE GLUTINEUSE • *Tofieldia glutinosa*

GENERAL: Perennial herb; stems 10–50 cm tall, covered in **sticky hairs**; from vertical rhizomes.

LEAVES: 2–4, basal, **linear**, 8–20 cm long, up to 8 mm wide; single, bract-like stem leaf.

FLOWERS: White, 6-parted, 4 mm wide; anthers purplish; arranged in a **dense cluster 2–5 cm long**; early summer.

FRUITS: Reddish or purplish capsules, 5–6 mm long.

WHERE FOUND: In rich fens and on Lake Superior shores; from Newfoundland to Alaska, south to California, Minnesota and Georgia.

NOTES: Least false asphodel (*T. pusilla*) is smaller and lacks the sticky stems. It is found from Quebec to Alaska, south to Lake Superior.

NODDING TRILLIUM
TRILLE PENCHÉ • *Trillium cernuum*

GENERAL: Low, perennial herb, less than 25 cm tall; from stout rhizomes.

LEAVES: 4–15 cm long, 6–10 cm wide, **diamond- or egg-shaped, hairless**, in a whorl of 3 at tip of stem.

FLOWERS: White (pale pink with age), 3-parted, with tips of petals and sepals bent back, **drooping below leaves on a long flexible stalk**; up to 5 cm wide; spring to early summer.

FRUITS: 6-lobed, red berries, with many seeds; early to mid-summer.

WHERE FOUND: In moist upland forests and black ash swamps and along riverbanks; from Newfoundland to Saskatchewan, south to Iowa, Alabama and Georgia.

NOTES: Trilliums are well known by herbalists. Native people believed that the roots, when chewed, provided an instant cure for rattlesnake bites. • The genus name *Trillium* means 'in 3s,' and it refers to the leaves, sepals and petals, which are in 3s. The species name *cernuum* means 'nodding.'

NORTHERN BLUE FLAG
IRIS VERSICOLORE • *Iris versicolor*

GENERAL: Perennial herb; stems 20–80 cm tall, in small colonies from **thick rhizomes**.

LEAVES: Linear, with parallel veins, up to 3 cm wide, **as long as stem** and sheathing stem near base.

FLOWERS: Showy, 6–8 cm wide, blue-purple, with **yellowish veins**; flower parts in 3s; large flower parts are sepals, inner, smaller 3 are petals.

FRUITS: Capsules, 3–5 cm long, with flat seeds stacked inside, turning dark brown in autumn (a useful identification feature after flowering).

WHERE FOUND: Along shores and in marshes, swamps, wet meadows and occasionally fens; from Newfoundland to southern Manitoba, south to Minnesota and Virginia.

NOTES: These flowers are pollinated by bees. Pollen from previously visited flowers is brushed from the bee's back as it enters to get nectar. • Muskrats, beavers and waterfowl eat the rhizomes, and ruby-throated hummingbirds feed on the nectar of the flowers. Irises are eaten by the iris borer, a moth whose larvae live in the leaves and eats their way to the rhizomes, killing the plant. Hollow rhizomes indicate that this parasite is present. • **The rootstocks are extremely poisonous.** Native people used the rhizomes as a strong cathartic. The rhizomes were taken after fasting to rid the body of disease. The Ojibway (and other North American people) used this plant as a charm against snakes. Parts of the plant were taken along and frequently handled when picking blueberries because it was believed that the scent of the plant would drive away snakes. • The genus name *Iris* is from the Greek meaning 'goddess of the rainbow,' and it refers to the many, brightly coloured flowers of these plants.

YELLOW LADY'S-SLIPPER
CYPRIPÈDE SOULIER • *Cypripedium calceolus*

GENERAL: Perennial herb, 20–80 cm tall; from coarse, fibrous roots.

LEAVES: Alternate, sheathing, oval, 5–15 cm long, half as wide as long, with parallel veins.

FLOWERS: Solitary (sometimes 2), **yellow, slipper-like**, at tip of stalk; **side petals greenish yellow to purplish brown, often twisted**; lip 2–6 cm long, inflated, pouch-shaped, yellow; **1 large green bract overtops flower**; spring and early summer.

FRUITS: Brown, 3-ribbed, persistent capsules, full of tiny seeds; ribs become slits through which seeds disperse as capsule blows about in wind; mid- to late summer.

WHERE FOUND: In calcium-rich fens, swamps, wet meadows and moist forests (conifer or mixedwood); from Newfoundland to the Yukon Territory, south to Oregon, Texas, Tennessee and Georgia.

NOTES: Showy lady's-slipper (*C. reginae*) is a large (up to 80 cm tall), leafy-stemmed lady's-slipper with white flowers tinged pink or rose-purple. The 3–4 cm long, lower lip of its flowers also has reddish to pinkish stripes. Showy lady's-slipper is locally common on calcium-rich soils in swamps, wet forests and fens, from Manitoba to Newfoundland, south to Missouri and Georgia. • Orchids have developed a symbiotic relationship with a fungus that helps the orchid absorb nutrients from the soil. • Bees enter the slipper-like pouch but cannot exit the same way. Once inside, they follow the passage to a lightly coloured exit with hairs covered by sticky nectar. The bees crawl over the hairs, past the stigma (depositing any pollen from previous flowers) and out the entrance, where gummy pollen is plastered on their hairy backs. • After germination, yellow lady's-slipper plants take several years to flower. • Native people used lady's-slipper plants to treat worms, kidney disorders in children, nervousness and various female disorders. They also wrapped it in sacred bundles, and used it to induce supernatural dreams. • The genus name *Cypripedium* is Greek, and it translates in English as 'Venus's slipper.'

C. reginae

DRAGON'S MOUTH
ARÉTHUSE BULBEUSE • *Arethusa bulbosa*

GENERAL: Showy, perennial herb,
12–20 cm tall; from small bulbous corms.

LEAVES: Solitary, grass-like, 10–15 cm long,
up to 7 mm wide, sheathing stems at base;
develop after plant blooms.

FLOWERS: Single, pink to
magenta, irregular, 3–5 cm
long, at stem tip above a pair
of small bracts; **lower lip has
3 indistinct lobes, spotted
purple and crested with
yellow hairs**; sepals erect and
curved to sides; late June to
mid-August.

FRUITS: Strongly 6-ribbed,
elliptic capsules, about 2.5 cm long, with hundreds of tiny seeds.

WHERE FOUND: In fens, often rooted in peat moss hummocks;
from Newfoundland to Saskatchewan, south to Indiana and New Jersey.

NOTES: Dragon's mouth is the only species of this genus in North America.
Another species is found in Japan. • Other pink fen orchids include **rose
pogonia** (*Pogonia ophioglossoides*, p. 62) and **swamp pink** (*Calopogon
tuberosus*, below) both of which have leaves at flowering time. Dragon's
mouth is generally the most common of these 3 species.

SWAMP PINK
CALOPOGON • *Calopogon tuberosus*

GENERAL: Perennial herb, 30–70 cm tall;
from swollen corms.

LEAVES: Single, grass-like, basal, up to 50 cm long
and 4 cm wide.

**FLOWERS: Several, showy,
bright magenta-pink, clustered at
tip of stem,** 20–25 mm in diameter;
**lip at top of flower, tipped with
yellow hairs;** early to mid-summer.

FRUITS: Oval capsules, with many
tiny seeds; mid- to late summer.

WHERE FOUND: In fens and conifer swamps, occasionally in moist
fields and along coast of Lake
Superior in 'boggy' beach pools;
from Newfoundland to southern Manitoba, south to Texas and Florida.

NOTES: Unlike most orchids, the flowers of swamp pink do not twist
180° during development. The lip, therefore, is at the top, rather than at
the bottom of the flower. • The genus name *Calopogon* means 'beauti-
fully haired or bearded,' and it refers to the yellow hairs on the pink
upper lip. An old name for this plant was *Limnodorum* which is Greek
for 'a meadow gift.'

ROSE POGONIA
POGONIE LANGUE-DE-SERPENT • *Pogonia ophioglossoides*

GENERAL: Perennial herb; stems 20–40 cm tall; from short rhizomes and fibrous roots.

LEAVES: Solitary, **oblong, 3–9 cm long, 1–2.5 cm wide, clasping stem near middle**.

FLOWERS: Solitary, 1.5 wide, 3 cm long, **pink with a yellow-bearded lip, above a single leaf-like bract**; early summer.

FRUITS: Capsules, oblong or egg-shaped, erect, full of seeds; mid-summer.

WHERE FOUND: In open fens and occasionally in swamps and wet fields and on lakeshores; from Newfoundland to northwestern Ontario, south to Texas and Florida.

NOTES: Rose pogonia grows in a habitat similar to that of **swamp pink** (*Calopogon tuberosus*, p. 62) and **dragon's mouth** (*Arethusa bulbosa*, p. 62). • The genus name *Pogonia* means 'bearded,' and refers to the lip of the orchid. The species name *ophioglossoides* means it looks like the fern *Ophioglossum*, which has leaves like a snake's tongue (from the Latin *ophio*, 'snake,' and *gloss*, 'tongue').

SMALL PURPLE FRINGED ORCHID
HABÉNAIRE PAPILLON • *Platanthera psycodes*

GENERAL: Striking perennial herb 15–90 cm tall; from a cluster of tuberous roots.

LEAVES: 2–5, keeled, elliptic or lance-shaped, 5–25 cm long; several smaller bracts also present.

FLOWERS: Pink-purple, with a fringed lip and slender spur; sepals thick, green; in a loose spike of 30–50 flowers; late July to early August.

FRUITS: Capsules, filled with many tiny seeds; late summer.

WHERE FOUND: In rich fens and wet meadows; from Newfoundland to southeastern Manitoba, south to Iowa, Tennessee and Georgia.

NOTES: This species is also known as *Habenaria psycodes*. • **Ragged orchid** (*P. lacera*, also known as *H. lacera*) has a fringed lip and white, cream to greenish-tinged, flowers. It is uncommon, from Newfoundland to Lake Superior, south to Texas and Florida. • **White-fringed orchid** (*P. blephariglottis*, also known as *H. blephariglottis*) is a similar, showy orchid with pure white flowers and narrower, lance-shaped leaves. Also, its flowers have simple (not 3-lobed), narrowly egg-shaped, fringed, lower lips. White-fringed orchid grows in wet, peaty soils, from Ontario to Newfoundland, south to Missouri and Florida. It is rare in Ontario, where it usually grows in open black spruce–tamarack fens, rooted in peat moss. • The species name *psycodes* means 'butterfly-like,' in reference to the colourful flowers.

P. blephariglottis

NORTHERN GREEN ORCHIS
HABÉNAIRE HYPERBORÉALE • *Platanthera hyperborea*

GENERAL: Robust perennial herb; **stems leafy**, 20–60 cm tall; from clusters of fleshy roots.

LEAVES: 3–6 on stem, 5–25 cm long and 3 cm wide, smaller near tip of stem.

FLOWERS: Greenish yellow, with a strap-shaped lip that is 4–7 mm long and **prolonged backward into spur**, not fringed; sepals thick, green; **in a compact spike 6–20 cm long**; mid-summer.

FRUITS: Capsules, filled with many tiny seeds; late summer.

WHERE FOUND: In ditches, fens and open swamps; from Newfoundland to Alaska, south to California, Nebraska, Ohio and New England.

NOTES: This species is also known as *Habenaria hyperborea*. • **Green woodland orchis** (*P. clavellata*, also known as *H. clavellata*) has several smaller leaf-like bracts and, near the base of the plant, a single leaf that is 5–15 cm long and approximately $^1/_4$ as wide. Its flowers are white, often tinged with green or yellow. Green woodland orchis is widespread from Newfoundland to Lake Superior, south to Texas and Florida, but it is less common than northern green orchis. • **Bog candle** (*P. dilatata*, also known as *H. dilatata*) has up to 12 stem leaves, which are smaller near the tip of the stem. Its flowers are fragrant (they smell like vanilla), white or tinged with green and they are arranged in a slender spike. The lip petal is expanded at the base. Bog candle is uncommon from Newfoundland to Alaska, south to California, South Dakota and New Jersey. Both of these species can hybridize with northern green orchis. • **White adder's mouth** (*Malaxis monophylla*) is a tiny (5–15 cm tall), slender orchid with solitary, 1.5–10 cm long leaves and narrow, elongate clusters (racemes) of many whitish flowers. Each flower is about 2 mm long, with a 3-lobed lower lip (at least 2–2.5 mm long) that is rounded at the base but tapered to a slender point at the tip. White adder's mouth is widespread but local in cold, wet soils of rich conifer swamps from Alaska to Newfoundland, south to California, Texas and New Jersey. This tiny plant is easily overlooked.

P. dilatata

Malaxis monophylla

P. hyperborea

P. clavellata

P. dilatata

HOODED LADIES'-TRESSES
SPIRANTHE PENCHÉE • *Spiranthes romanzoffiana*

GENERAL: Perennial **herb**; stems **erect**, 15–40 cm tall (rare to 60 cm), usually hairy near tip and bearing 4–5 small bracts; from slender, fleshy roots.

LEAVES: Mainly **basal**, **5–10 mm wide**, often slightly wider towards tip, 7–35 cm long, sometimes noticeably stalked.

FLOWERS: Yellowish white, 7–10 mm long, spreading or nodding, with a strong, almond- or vanilla-like fragrance; lower lip oblong-egg-shaped, with wavy or torn edges; arranged **in 2–3 rows in dense, spirally twisted clusters** (racemes) 10–12 cm long and 12–15 mm wide.

FRUITS: Ovoid to oblong, erect capsules.

WHERE FOUND: In many habitats, including fens, lakeshores, stream-banks, pools, marshes, roadside ditches and wet meadows; occasionally in swamps, damp woods, fields and old roadbeds; most common in wet, marly and fairly open conditions, often associated with horsetails (*Equisetum* spp.) and arrowgrasses (*Triglochin* spp.); from Alaska to Newfoundland, south to California, New Mexico, Ohio and New England; this North American species has relict populations in Ireland and Scotland.

NOTES: Hooded ladies'-tresses could be confused with **nodding ladies'-tresses** (*S. cernua*) but the side sepals of that species are mostly 2–3 mm wide, its petals and sepals are only faintly veined, and its lower lip is not distinctly constricted. Nodding ladies'-tresses blooms in late summer and early autumn. It sometimes grows with (and hybridizes with) hooded ladies'-tresses, but it is also found in somewhat drier conditions. Habitats include moist, open meadows, roadside ditches, recent excavations, lakeshores, riverbanks, sand and gravel flats and occasionally swamps, from Minnesota to Nova Scotia, south to Texas and Florida. • Some native peoples used hooded ladies'-tresses to treat venereal diseases and urinary disorders. • Hooded ladies'-tresses is known as *beemsquandawish* in Ojibway. The genus name *Spiranthes*, is derived from the Greek *speira*, 'a coil or spiral,' and *anthos*, 'a flower,' in reference to the spirally twisted flower clusters. This species was named after Count Nicholas Romanoff, an 18th century Russian minister of state and patron of the sciences who first described the species. The species name of nodding ladies'-tresses, *cernua*, is Latin for 'face-down,' in reference to the frequently nodding flowers.

S. cernua

HEART-LEAVED TWAYBLADE
LISTÈRE CORDÉE • *Listera cordata*

GENERAL: Small, delicate perennial herb; stems hairless, 10–25 cm tall; from slender rhizomes.

LEAVES: 2, opposite, heart-shaped, resembling a pair of ears midway up flowering stem, 1–3 cm long, stalkless.

FLOWERS: Pale purplish green, 5–7 mm wide; lip 3–6 mm long, deeply cut, with 2 slender lobes at tip and 2 teeth at base, twice as long as side petals; in a spike at stem tip; early to mid-summer.

FRUITS: Small brown capsules, full of tiny seeds; mid- to late summer.

WHERE FOUND: In conifer swamps and moist conifer forests; often overlooked as it blends into carpets of moss; from Newfoundland to Alaska, south to California, Minnesota and North Carolina.

NOTES: Heart-leaved twayblade is the most common twayblade in northern Ontario. Most other twayblades are uncommon and have flowers with shallowly cut lips and broad lobes. • The lower lips of the flowers of **southern twayblade** (**L. australis**) are also slender and forked, but they lack the 2 basal teeth found on flowers of heart-leaved twayblade, and they are 4–8 times as long as the side petals. Southern twayblade grows in damp woods, thickets and bogs from Ontario to Nova Scotia, south to Louisiana and Florida. It is rare in Ontario, where it grows in black spruce–tamarack *Sphagnum* peatlands. • **Broad-lipped twayblade** (**L. convallarioides**) has larger flowers with a broad, wedge-shaped, 9–10 mm long lip ending in 2 broad, rounded lobes. It is also uncommon, but it grows in cedar swamps and on the edges of peatlands from Alaska to Newfoundland, south to California, Michigan and Tennessee. • The species name *cordata* means 'heart-shaped,' in reference to the leaves. The common name 'twayblade' refers to the 2 opposite leaves.

L. australis

L. convallarioides

WOOD NETTLE, CANADA NETTLE
LAPORTÉA DU CANADA • *Laportea canadensis*

GENERAL: Perennial **herb** with **stinging hairs**; stems erect or ascending, 45–90 cm tall (rarely to 120 cm).

LEAVES: Alternate, egg-shaped, 7–17 cm long, 5–13 cm wide, thin, strongly feather-veined, **long-stalked, sharply toothed**.

FLOWERS: Small, **green**, lacking petals, male or female, with both sexes on same or separate plants; **in large, spreading, loose clusters** (cymes) from leaf axils; male flowers 5-parted, in lower clusters; female flowers 4-parted, in upper clusters.

FRUITS: Hairless achenes, flat, **egg-shaped**, about 2.5 mm long, oblique, **bent backward** on their stalks.

WHERE FOUND: In hardwood swamps and damp woods and along streambanks, preferring rich, alluvial soil and often found in old lumber camp clearings, usually on old manure or compost dumps; from southeastern Saskatchewan to Nova Scotia, south to Oklahoma and Florida.

NOTES: Slender stinging nettle (*Urtica dioica* ssp. *gracilis*, also known as *U. gracilis*) is another common plant with stinging hairs on its leaves and stems. However, slender stinging nettle has opposite leaves, all of its flowers are 4-parted, and its achenes are straight and erect. It grows on moist, nitrogen-rich soil in damp woods, barnyards and compost piles, along streams, rivers and roadsides and on islands with bird nesting colonies, from Alaska to Labrador, south through the U.S. and Mexico to South America. • **False nettle** (*Boehmeria cylindrica*) and **clearweed** (*Pilea pumila*) are 2 members of the nettle family that lack stinging hairs. Both have opposite, long-stalked, coarsely toothed, egg- to lance-shaped leaves, but their flower clusters are very different. The flowers of false nettle are borne in dense, interrupted spikes that are often longer than the leaf stalks and may be tipped with a pair of leaves. Clearweed flowers are borne in smaller, less dense clusters that are much shorter than the leaf stalks. False nettle is found in swamps and mixedwoods from Minnesota to Maine, south to Texas and Florida. Clearweed is found in moist, rich soil in cool or shaded locations from Ontario to Prince Edward Island, south to Texas and Florida. • The bristly hairs of wood nettle and slender stinging nettle contain an irritating oil rich in **formic acid**. When the plant is touched, the hairs act like minute hypodermic needles, releasing this oil into the skin and causing **painful stinging**. The reddness, swelling, burning and itching that follow can last for several minutes to a few hours. • The stinging properties of nettles are eliminated by cooking, and young, tender nettle leaves make a delicious, protein-rich potherb. The cooking water can be used as soup stock or tea concentrate, and dried leaves can be used throughout the year to make a pleasant, nutritious tea that is rich in iron and vitamin C. • Wood nettle is also used for fibre and dye. In Russia, good yields of high quality 'hay' are obtained from stinging nettle. Wood nettle pollen is shed in large amounts and contributes to summer hay fever. • This genus was named in honour of Francois L. de Laporte, the Count of Castelnau, a 19th century entomologist.

Urtica dioica

LADY'S THUMB
RENOUÉE PERSICAIRE • *Polygonum persicaria*

GENERAL: Large, annual herb; stems simple or branched, 30–50 cm tall, erect to ascending; from stout taproot.

LEAVES: Narrowly lance-shaped, 1–10 cm long, **green, with a purplish blotch near centre**, short-stalked; **sheaths** (ocreae) **loose**, minutely hairy to nearly hairless, **fringed** with short bristles.

FLOWERS: Pink to reddish, small; in **dense spikes, 1–5 cm long**, oblong, stalked, erect, solitary or in branching clusters; June–July.

FRUITS: Black, shiny achenes, 2–2.5 mm long, lens-shaped or 3-angled; September–October.

WHERE FOUND: In wet areas and waste places from Alaska to Newfoundland and south through the U.S.; introduced from Eurasia.

NOTES: Other *Polygonum* species (except **water smartweed** [*P. amphibium*, p.168]) have bristle-fringed ocreae. • **Carey's knotweed** (*P. careyi*) has nodding, pink to purplish spikes and rough-glandular stems, branches and flower stalks. It grows on shores and in fields and disturbed areas from Minnesota to southwestern Quebec, south to Indiana and Delaware. It is rare in Ontario. • **Marsh pepper smartweed** (*P. hydropiper*) and **dotted smartweed** (*P. punctatum*) both have dark, glandular dots on their sepals, but marsh pepper smartweed has dull, granular achenes and nodding spikes, whereas dotted smartweed has shiny achenes and erect spikes. Both are found across North America in swamps, marshes and ditches. • The smartweeds are highly variable plants that can take on very different growth forms (phases) under different environmental conditions. Submerged plants are usually hairless and broad-leaved, plants growing out of water are usually more or less hairy and narrow-leaved, and plants growing on wet, muddy sites are intermediate in character.

NORTHERN COMANDRA
COMANDRE LIVIDE • *Geocaulon lividum*

GENERAL: Erect, perennial herb, 10–30 cm tall; from **slender, red rhizomes**.

LEAVES: Alternate, thin, hairless, 1–4 cm long, entire, **bright green (or mottled with yellow)** and rounded at tip.

FLOWERS: 2 mm in diameter, in **clusters** of 2–4 from **leaf axils**; **5 greenish inconspicuous tepals**; early to mid-summer.

FRUITS: Orange berries, 5–10 mm in diameter, 1-seeded; late summer to early autumn.

WHERE FOUND: In black spruce swamps, treed fens, bogs and upland forests (feather moss habitats); from Newfoundland to Alaska, south to Washington, Minnesota and New England.

NOTES: Northern comandra may be confused with **bastard toad-flax** (*Comandra umbellata*), but that species is found in dry uplands and has flowers in clusters at the stem tip. • The fruit is said to be edible, but it is not very tasty. The leaves are sweet-tasting and have been used in salads. • Northern comandra is parasitic on the roots of ericaceous shrubs. • The genus name *Geocaulon* means 'stem that is found under the ground,' and it refers to the slender red rhizomes.

PENNSYLVANIA BITTERCRESS
CARDAMINE DE PENNSYLVANIE • *Cardamine pensylvanica*

GENERAL: Annual or biennial herb; flowering stems **leafy**, 10–50 cm tall, erect or spreading; from taproots.

LEAVES: Compound, with **3–5 pairs of oblong to lance-shaped leaflets** and a **larger leaflet at tip,** not usually found in large masses.

FLOWERS: White, 4-parted, about 4 mm in diameter.

FRUITS: Erect, linear pods (siliques) 1–3 cm long, 0.7–1 mm wide, on ascending stalks (pedicels), curled after discharge of seeds; 24–40 seeds, 1–1.5 mm long.

WHERE FOUND: On wet, muddy shorelines, streams, riverbanks and beaver dams and in rich swamps and low, wet woods; from Alaska to Newfoundland, south to California, Texas and Florida.

NOTES: Pennsylvania bittercress could be mistaken for **watercress** (*Nasturtium officinale*, also known as *Rorippa nasturtium-aquaticum*) which has white flowers, fibrous roots, coarsely reticulated seeds and sometimes hollow stems. Watercress can be submerged, partly floating, erect or sprawling on mud. It can usually be distinguished by the adventitious roots along its stems and by its widely spreading flowers. Also, the valves of watercress fruits are more rounded than those of Pennsylvania bittercress, and they do not curl at maturity. Watercress was introduced from Europe. It grows in rich deciduous swamps, streams and riverbanks from Alaska to Newfoundland, south through much of the U.S. • The young shoots make delicious salad greens, and watercress is used in mixed juices, soups, stews and casseroles and as a garnish. • Ducks and muskrats eat watercress and other cress species. • There are several possible origins of the genus name *Cardamine*. It may have come from *kardamon*, the Greek name for a mustard (later applied to the bittercresses), or it may have come from the Greek *kardia* , 'heart,' and *damao*, 'to overpower or to calm,' in reference to the use of these plants as a heart medicine, poison or sedative. The species name *pensylvanica* refers to Pennsylvania, where the first specimens originated.

Nasturtium officinale

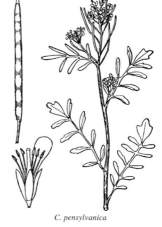

C. pensylvanica

MARSH MARIGOLD
POPULAGE DES MARAIS • *Caltha palustris*

GENERAL: Early spring, perennial herb, 20–60 cm tall; stems hollow and succulent; often loosely **clumped** from coarse roots.

LEAVES: Broadly heart-shaped, 1.5–4 cm wide; **basal leaves long-stalked**; **stem leaves short-stalked**.

FLOWERS: Bright yellow, 1.5–4 cm wide, with 5–9 petal-like sepals and no petals; in clusters at stem tip; spring to early summer.

FRUITS: Recurved capsules (follicles), splitting open along inside wall, many-seeded; early to mid-summer.

WHERE FOUND: In swamps, marshes and wet ditches and on streambanks and lakeshores; from Newfoundland to Alaska, south to Oregon, Nebraska and South Carolina.

NOTES: Marsh marigold is one of the first plants to flower in early spring. • All parts of the fresh plants contain protoanemonin and helleborin, **poisons** that are toxic to the heart and cause inflammation of the stomach. The **leaves can even cause skin to blister**. In the past, poultices of the leaves were used as counter-irritants to relieve rheumatic pain, and the caustic juice was dripped onto warts. The poisonous principle is volatile, however, so cooked or dried plants are harmless. • Grouse eat the seeds, and deer and moose eat the plants.

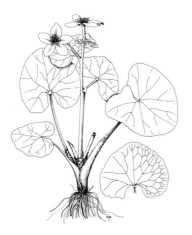

CANADA ANEMONE
ANÉMONE DU CANADA • *Anemone canadensis*

GENERAL: Perennial herb; flowering stems hairy, 30–60 cm tall, **branching at whorl** of stem leaves; often forming large patches.

LEAVES: Basal and on the stem; basal leaves stalked, with 3–7 **broadly wedge-shaped, toothed divisions**; **stem leaves stalkless**, 2–3-parted, **in whorls** below flowers

FLOWERS: White, 2.5–3 cm wide, with **5 petal-like sepals**.

FRUITS: Stalkless, flattened, **woolly achenes**, tipped with a straight bristle (style); in dense, rounded heads.

WHERE FOUND: In rich swamps, floodplains, sandy shores, thickets, wet meadows, roadsides and drier habitats; from British Columbia and southern Northwest Territories to Nova Scotia, south to New Mexico, Missouri and New Jersey.

NOTES: **Wood anemone** (*A. quinquefolia*) also has a solitary basal leaf arising from a perennial rhizome, but it is usually only 10–20 cm tall. Flowering plants have a whorl of 3 stem leaves that are smaller (2–5 cm long) and distinctly stalked, and which are often cut so deeply as to appear to be 5 leaflets. Wood anemome grows in many habitats, including swamps and moist woods, from British Columbia to Nova Scotia, south to northern California, Iowa and North Carolina. • Wood anemone flowers close at night, for protection, and they open during the day when their pollinators are active. Once pollinated, they remain closed. • Anemones contain a **toxic** compound (anemonine), which can irritate the skin. • Native peoples used ground anemone roots to make a poultice to treat wounds and bleeding. Some native groups also used it to clear the throat before singing. A tea made from the roots was believed to help soothe lung diseases. • The Ojibway name for Canada anemone is *midewidjeebik*, which means 'echo root.' Anemones have been called 'wind flowers,' and 'anemone' comes from the Greek word *anemos*, 'wind.' This may reflect the old belief that anemones bloomed only when the wind blew, or that wind that had passed over a field of anemones was poisoned, and that disease followed in its wake. Also, anemones often grow in open, wind-swept areas, where their slender stalks tremble in a breeze.

GOLDTHREAD, CANKER-ROOT
COPTID DU GROENLAND • *Coptis trifolia*

GENERAL: Perennial herb, 5–15 cm tall; from **slender, golden rhizomes**.

LEAVES: Evergreen, leathery, shiny, basal, 2–5 cm wide, compound; 3 leaflets, obscurely-lobed and toothed.

FLOWERS: White, 12–16 mm wide, star-shaped, with 5–7 petal-like sepals, solitary on leafless stalks; early to mid-summer.

FRUITS: Several seed-filled capsules in a whorl or loose cluster; mid- to late summer.

WHERE FOUND: In conifer swamps and moist upland forests; from Newfoundland to Alaska, south to southern British Columbia, Iowa and North Carolina.

NOTES: Grouse feed on these fruits. • Native people used the roots to make a soothing tea for teething babies, an eye wash and a bright yellow dye. • The species name *trifolia* means '3 leaflets.' 'Canker-root' comes from its traditional use for treating ulcers of the mouth, throat and stomach.

TALL MEADOW-RUE
PIGAMON PUBESCENT • *Thalictrum pubescens*

GENERAL: Stout, perennial herb, usually **purplish**; stems **erect**, 0.5–2.5 m tall, **branching, leafy**.

LEAVES: Compound, **3–4 times divided in 3s**; leaflets light green, paler beneath, oblong to rounded, **with 3 main, pointed lobes**.

FLOWERS: White (rarely purplish), male and/or female, with unisexual and bisexual flowers in the same plant; **in compound, leafy clusters (panicles)** 30 cm long or more; anthers 0.8–1.5 mm long, **filaments club-shaped** at tips, often broader than anthers.

FRUITS: Ovoid, short-stalked **achenes, with 6–8 winged angles**, narrowing to the base.

WHERE FOUND: On damp streambanks, lakeshores and floodplains and in wet meadows, thickets, marshes and open swamps; from Ontario to Newfoundland, south to Tennessee and Georgia.

NOTES: This species was previously known as *T. polygamum*. • Purple meadow-rue (*T. dasycarpum*) is found in similar habitats and could be confused with tall meadow-rue (some classify these two as a single species). Purple meadow-rue can be distinguished by its longer anthers (1.5–3.5 mm), with slender or only slightly club-shaped filaments (narrower than the anther) and darker green leaves. It grows in meadows, damp thickets and swampy areas from southern British Columbia to western Quebec, south to Arizona, Louisiana and Ohio. • Many native peoples considered meadow-rue to be a powerful love medicine. Some hid the seeds in the food of quarrelling couples to help love overcome their differences. • The species name *pubescens* means 'densely hairy.'

PITCHER-PLANT
SARRACÉNIE POURPRE • *Sarracenia purpurea*

GENERAL: Unmistakable, insectivorous, perennial herb, with tubular, hollow leaves; from a stout rootstock.

LEAVES: In basal rosettes, 10–20 cm tall, **pitcher-like**, reddish green, with a narrow wing on upper side and a hood over opening at tip, hollow, with **downward-pointing hairs inside**, holding water and entrapping insects.

FLOWERS: Purple to dark red or maroon, **5–7 cm wide**, 5-parted with 5 large sepals and 5 incurved maroon petals, **nodding above leaves on 3–50 cm tall stalks**; early to mid-summer.

FRUITS: Capsules, with many small seeds; mid- to late summer.

WHERE FOUND: In bogs, fens and open conifer swamps; from Newfoundland to the Northwest Territories, south to southern Saskatchewan, Minnesota, Ohio and Delaware.

NOTES: Insects are attracted into these leaves by the reddish colour and musty odour. They become trapped because they cannot climb out over the downward-pointing hairs. The plant releases a wetting agent and digestives enzymes to dissolve its prey. It then absorbs nitrogen and other nutrients from the liquid. • The larvae of some moths burrow into the pitcher-plant and feed on it. The larva of a small mosquito, *Wyeomyia smithii*, lives only in the water of pitcher-plant leaves. • Native people used this plant as a stimulant and diuretic, to hasten childbirth and to treat dysentery. • The name *purpurea* means 'purple' and refers to the purplish leaves and flowers. Linnaeus named this plant after Sarrazin, a Quebec physician who used it to treat smallpox in the early 1700s.

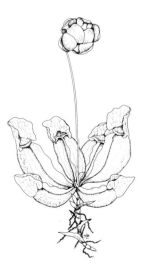

ROUND-LEAVED SUNDEW
ROSSOLIS À FEUILLES RONDES • *Drosera rotundifolia*

GENERAL: Small, insectivorous, perennial herb; from fibrous roots.

LEAVES: In a **basal rosette** up to 10 cm wide, covered with **reddish, sticky gland-tipped hairs**; blades rounded, **wider than long.**

FLOWERS: White (rarely pink), **in 1-sided, elongated clusters (racemes)**; 5 sepals; 5 petals, 4–6 mm long; early to mid-summer.

FRUITS: Capsules, with 3–5 compartments and many seeds; mid- to late summer.

WHERE FOUND: In bogs, fens, swamps (black spruce–tamarack and eastern white cedar) and in mossy crevices along lakeshores; from Newfoundland to Alaska, south to California, Minnesota and Florida.

NOTES: **Spatulate-leaved sundew** (*D. intermedia*) has spatula-shaped leaves that are less than 6.5 times longer than wide, and that are widest near their tips. It is common in fens and swamps and along lakeshores, from Newfoundland to Lake Superior, south to Texas and Florida. • **Linear-leaved sundew** (*D. linearis*) has even narrower, linear leaves that are at least 7 times longer than wide, with parallel edges. It grows in rich fens, from Newfoundland to British Columbia, south to Minnesota, southern Ontario and Maine. • **English sundew** (*D. anglica*) has elongate leaves with non-parallel edges. It grows in rich fens from Newfoundland to British Columbia, south to Minnesota, southern Ontario and Maine. • Sundew leaves are equipped with hairs tipped with sticky droplets that adhere to insects, entrapping them. As an insect struggles, it touches more sticky glands and stimulates the leaf to secrete digestive enzymes that pool up around the insect. These enzymes digest the insect and the nutrients accumulate in the fluid. Once the insect is digested, the secretion pool with all its newly acquired nutrients is absorbed into the leaf, and the nutrients are distributed throughout the plant for growth and development. • Traditionally sundews have been used as a remedy for whooping cough and asthma. The juice from the leaves has been used to eliminate pimples and corns. • The genus name *Drosera* means 'dew,' and it refers to the sticky droplets on the leaves.

D. linearis

NAKED MITREWORT
MITRELLE NUE • *Mitella nuda*

GENERAL: Low, perennial herb, up to 20 cm tall; stems leafless, hairy; from spreading rhizomes.

LEAVES: Basal, rounded to kidney-shaped, 2–5 cm wide, palmately veined, blunt-toothed; **upper surface with distinctive, erect hairs**.

FLOWERS: Whitish green, snowflake-like, with 5 sepals and **5 fringed petals,** stalked, in an elongated cluster (raceme); spring to early summer.

FRUITS: Cup-like capsules, open widely at tip, with few, shiny black seeds; mid- to late summer.

WHERE FOUND: In moist upland forests and swamps and along shaded streambanks; an indicator of relatively rich conditions; from Newfoundland to Alaska, south to Montana, Minnesota and Pennsylvania.

NOTES: Mitreworts without flowers might be confused with **foam flower** (*Tiarella cordifolia*), which has 5–10 cm long leaves and tiny, simple petals, and its capsules split into 2 unequal parts. It grows in drier, rich upland forests, from Nova Scotia to central Ontario, south to Tennessee and North Carolina. • Grouse occasionally eat the seeds of naked mitrewort. • The genus name *Mitella* means 'little mitre,' and it refers to the shape of the young fruit, which looks like a tall hat (mitre) worn by bishops and other ecclesiastics.

M. nuda

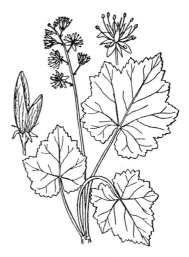

Tiarella cordifolia

MARSH GRASS-OF-PARNASSUS, NORTHERN GRASS-OF-PARNASSUS
PARNASSIE PALUSTRE • *Parnassia palustris*

GENERAL: Hairless, perennial herb; stems erect, 5–30 cm tall, **with a single, clasping, heart-shaped leaf** (usually below middle) from short rhizomes.

LEAVES: Basal, firm, **heart-shaped**, 20–40 mm long, on long, slender stalks.

FLOWERS: White, showy, 15–25 mm wide, with **5 faintly veined petals**; 10 stamens, 5 fertile stamens alternating with sterile stamens, **sterile stamens with 9–15 slender, gland-tipped segments**.

FRUITS: Egg-shaped capsules, 8–12 mm long, containing many seeds.

WHERE FOUND: In marshes, shoreline meadows, fens and ditches, usually associated with calcium-rich soils; from Alaska to Newfoundland, south to California and Michigan.

NOTES: Carolina grass-of-Parnassus (**P. *glauca*,** previously known as *P. caroliniana*) is a larger (15–60 cm tall) plant, with showy, 20–40 mm wide, white flowers. Its sterile stamens are usually divided into 3 stout segments (staminodia), and they are shorter than the fertile stamens. It is often found with marsh grass-of-Parnassus. Habitats include wet meadows, marshes, calcareous sandy or gravelly shores, interdunal flats, streambanks, alkaline fens and occasionally the edges of swamps and forests, from Saskatchewan to Newfoundland, south to South Dakota, Illinois and New England. • **Small-flowered grass-of-Parnassus** (*P. parviflora*) is very similar to marsh grass-of-Parnassus (it is sometimes considered a subspecies), but its flowers are smaller (8–10 mm wide), and its leaf bases gradually taper to the stem (they are not heart-shaped). It is often found with **white spruce (*Picea glauca*, p. 15)** in damp, calcareous, sandy shores, meadows, interdunal flats, excavations, bluffs and rocks, from northern British Columbia to Newfoundland, south to Idaho, Wisconsin and southern Quebec. • The flowers of marsh grass-of-Parnassus have lines on the petals that act as 'runway lights' to lead insect pollinators to the attractive, nectar-tipped staminodia. • These plants were first named by Dioscorides, in honour of Mount Parnassus in Greece. The plants were probably dedicated or sacred to the Muses who were believed to live on the snow-capped mountain. The species name *palustris* means 'of the marsh or swamp.'

P. glauca

Rose Family (Rosaceae)

COMMON STRAWBERRY
FRAISIER DES CHAMPS • *Fragaria virginiana*

GENERAL: Low, perennial herb; from **thick rhizomes** and **reddish stolons**.

LEAVES: Basal, compound, with **3 leaflets, toothed; tooth at tip is shorter than 2 adjacent teeth.**

FLOWERS: White, up to 2 cm wide, **5-parted**, with 5 sepals and 5 petals; in loose clusters that are **shorter than leaves**; early summer.

FRUITS: Fragrant, red berries, with seeds embedded in pits on surface; mid-summer.

WHERE FOUND: In swamps and on lakeshores; often forming patches in ditches along roads, in forests, in open fields and in many other upland habitats; from Newfoundland to Alaska, south to California, Oklahoma, Tennessee and Georgia.

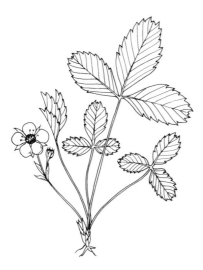

NOTES: Wood strawberry (*F. vesca*) has leaflets on which the tooth at the tip is **longer** than the adjacent teeth. Also, its flowers extend above the leaves and its fruits are more elongated. It is found from Newfoundland to Alaska, south to California, Nebraska, Illinois and Virginia. • Strawberries are a favourite food of many birds. Grouse eat the young leaves and ripe berries. Crows, catbirds, grosbeaks, sparrows, thrashers, waxwings and robins are a few of the songbirds that eat the berries. Hares, rabbits, skunks, squirrels, chipmunks, mice, deer and bears eat the fruits and leaves. • These berries are small but delicious, and they can be made into juices, jams and jellies. • Common strawberry was called *odaemin* by the Ojibway, who dried it and used it as a winter staple. The leaves were used to make tea and the fruit was used in cooking. Common strawberry was also used to treat various medical problems, such as stomach aches. Strawberries are known throughout the world as blood purifying and strengthening herbs.

F. vesca

MARSH CINQUEFOIL
POTENTILLE PALUSTRE • *Potentilla palustris*

GENERAL: Perennial herb, 20–60 cm tall; **stems red, erect, from woody rhizomes that trail across water or mud.**

LEAVES: Pinnately compound, with 5–7 coarsely-toothed leaflets, each 5–10 cm long and 1–3 cm wide; stalks long, with **2 sheathing stipules** at base.

FLOWERS: Few, reddish purple, 5-parted, with petals shorter and narrower than sepals, approximately 2 cm wide; throughout summer.

FRUITS: Achenes, 1 mm long, red to golden brown, hairless; mid- to late summer.

WHERE FOUND: In marshes and fens and on streambanks; from Newfoundland to Alaska, south to California, Iowa, Pennsylvania and New Jersey.

NOTES: Other cinquefoils have yellow or white flowers. • Another purple-flowered member of the rose family is **purple avens** (*Geum rivale*), also called **water avens**. The leaves of purple avens are pinnately compound, with 1–3 large, broad leaflets at the tip and several smaller leaflets below, often interspersed with tiny leaflets. Purple avens flowers have a bell-shaped calyx and do not open widely. They are borne in few-flowered, nodding clusters that become erect in fruit. The achenes of purple avens are hairy and have a 6–8 mm long, jointed style with a bend at the middle. This plant grows in fens, swamps and wet meadows from British Columbia to Newfoundland, south to Washington, New Mexico, Indiana and New Jersey. • Grouse eat marsh cinquefoil seeds. Rabbits and hares occasionally eat the leaves. • The genus name *Potentilla* means 'powerful herb,' and it refers to the use of some species as medicinal plants. Cinquefoils have been used as medicines throughout North America, Europe and Asia. The species name *palustris* means 'of the marsh or swamp.'

Geum rivale

CANADA ST. JOHN'S-WORT
MILLEPERTUIS ASCYRON • *Hypericum canadense*

GENERAL: Annual or perennial herb; stems erect, 10–40 cm tall, with **angular** branches; from fibrous roots.

LEAVES: Opposite, 3-nerved, blackish- or purplish-dotted with glands, **linear**, 1–4 cm long, 1–6 mm wide, blunt-tipped.

FLOWERS: Orange-yellow, 5-parted, 4–6 mm in diameter, several to many in **branching clusters** (cymes).

FRUITS: Red or purplish, narrowly cone-shaped, **pointed capsules**, 4–8 mm long, 1-celled, with many striped seeds.

WHERE FOUND: On damp shores of beaver ponds and lakes, on peatland edges and in disturbed areas; from southern Manitoba to Newfoundland, south to Iowa and Georgia.

NOTES: Three other species of St. John's-worts are also found in fens and on shorelines. • **Northern St. John's-wort** (*H. mutilum*) has much wider (egg- to lance-shaped), 5–7-nerved leaves and smaller (2.5–3.5 mm long), pointed capsules. It grows along the edges of rich beaver ponds, streams, meadows and marshes from Manitoba to Nova Scotia, south to Texas and Florida. • **Larger Canada St. John's-wort** (*H. majus*) has slightly wider (lance-shaped), 5–7-nerved leaves and large (6–8 mm long), blunt-tipped capsules. It grows along damp sandy to mucky shores, borders of streams and cedar swamps and in marshy or sedgy areas from southern British Columbia to Nova Scotia, south to Washington, Nebraska and Delaware. • **Pale St. John's-wort** (*H. ellipticum*) has fewer, larger (1–1.5 cm in diameter), pale yellow flowers. It grows on damp sandy shores, streambanks and peatland edges and in open thickets from Lake Superior to Newfoundland, south to Iowa and Maryland. • Some St. John's-worts are eaten by ducks. • The common name may have arisen from the blooming period, which is around June 24th, St. John's Day.

H. mutilum

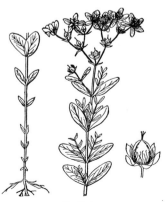

H. ellipticum

H. majus

MARSH ST. JOHN'S-WORT
MILLEPERTUIS DE FRASER • *Triadenum fraseri*

GENERAL: Perennial herb; stems 30–60 cm tall; from rhizomes.

LEAVES: Opposite, 3–6 cm long, egg-shaped, smooth, not toothed, **often with black dots (glands) beneath**, turning a distinctive purplish colour in late summer.

FLOWERS: Purple or flesh-coloured, 1–2 cm wide, with 5 petals and 5 pointed sepals (3–5 mm long); styles 0.5–1.5 mm long; in clusters at stem tips or in leaf axils; mid-summer.

FRUITS: Capsules with 3 compartments containing brown seeds; late summer to early autumn.

WHERE FOUND: On peaty shores and in floating fens and beaver meadows; from Newfoundland to Saskatchewan, south to Nebraska, Alabama and Georgia.

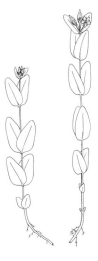

NOTES: Also known as *Hypericum virginicum* var. *fraseri*. • Four other species of St. John's-wort with yellow-orange flowers are found in fens, shorelines and wet meadows (see p. 78). • **Virginia St. John's-wort** (*T. virginicum,* previously known as *H. virginicum* var. *virginicum*) is very similar to marsh St. John's-wort, but its flowers are larger, with 5–8 mm long, obtuse or rounded sepals and 2–3 mm long styles. It is found in the same habitats as marsh St. John's-wort from southern Ontario to Nova Scotia, south to Georgia. • These flowers are usually closed in full sun and actually seem to be in bud throughout the summer. The large, purple-green flower buds are a good identification feature. • Native people used marsh St. John's-wort to make a tea to treat fevers.

T. virginicum

79

JEWELWEED, TOUCH-ME-NOT
IMPATIENTE BIFLORE • *Impatiens capensis*

GENERAL: Succulent, annual herb; erect, 50–150 cm tall; **stems brittle, juicy, reddish**; from fibrous roots.

LEAVES: Alternate, 3–10 mm long, **egg-shaped, coarsely toothed, pale, soft**.

FLOWERS: Irregular, **orange with reddish-brown spots**, 2–3 cm long, with a **short inward curved spur at back, hanging by slender stalks**; throughout summer.

FRUITS: Long-ribbed pod-like capsules, **2 cm long, green; mature pods explode when touched**, ejecting seeds; mid-summer to early autumn.

WHERE FOUND: In swamps, wet fields, shores, ditches and marshes; from Newfoundland to the Northwest Territories, south to southern British Columbia, Minnesota, Alabama and Florida.

**NOTES: Yellow jewelweed (*I. pallida*) is an uncommon, calcium-loving plant found from Newfoundland to southern Ontario, south to Kansas, Tennessee and Georgia. It grows in moist, shaded areas and has yellow flowers with a shorter spur. • Jewelweed flowers are an important source of nectar for hummingbirds. Grouse and mice eat the fruits, and hares and rabbits eat the stems and leaves. • The explosive fruits burst at the slightest touch when ripe. • This plant was used by native people as a remedy for stinging nettles and poison ivy, and to produce an orange or yellow dye.

AMERICAN DOG VIOLET
VIOLETTE DÉCOMBANTE • *Viola conspersa*

GENERAL: Hairless, perennial herb; **stems leafy**, 8–15 cm tall; from oblique, often branched rhizomes.

LEAVES: Alternate, rounded, **heart-shaped, rounded at tip**, 2–3 cm wide; **stipules** lance-shaped, **fringed** with sharp, slender-tipped teeth.

FLOWERS: From leaf axils, nodding, with 5 unequal petals, of 2 types; showy spring flowers numerous, **pale violet** (sometimes white), raised **above leaves**, lowest petal with a spur (nectary), April–May; inconspicuous summer flowers short-stalked, usually lacking petals, self-fertilizing (cleistogamous).

FRUITS: Capsules, splitting opening and projecting light brown seeds.

WHERE FOUND: Widespread and common on low, shaded ground; from southern Manitoba to Nova Scotia, south to Alabama and Georgia.

NOTES: The leaves of violets can be cooked like spinach and the flowers can be made into jams, jellies or syrup. Violets are very high in vitamins A and C. They contain 3 times more vitamin C by weight than do oranges! • Grouse and juncos eat violet fruits, and hares, rabbits and mice feed on the seeds and leaves. The leaves are a preferred food for many insect larvae and slugs. The flowers are pollinated by bees, which feed on the nectar, and the seeds are dispersed by ants.

WOOLLY BLUE VIOLET
VIOLETTE PARENTE • *Viola sororia*

GENERAL: Perennial herb, **lacking leafy stems**; from short, thick rhizomes.

LEAVES: Alternate, **heart-shaped, blunt-tipped**, often 10–12 cm wide, round-toothed, **silky hairy** on stalks and lower surface (especially when young), growing directly **from rhizomes**.

FLOWERS: From rhizome, nodding, with 5 unequal petals, of 2 types; showy spring flowers **violet to lavender** (occasionally white), their lowest petal essentially hairless and spurred; inconspicuous summer flowers ovoid, self-fertilizing (cleistogamous), on short, horizontal stalks, usually underground.

FRUITS: Capsules, 8–12 mm long, usually purple mottled with brown, on erect stalks, splitting open and projecting dark brown seeds.

WHERE FOUND: Moist, shady sites; from southern Manitoba to southwestern Quebec, south to Oklahoma and North Carolina.

NOTES: Two similar, but hairless violets of moist to wet habitats are **marsh blue violet** (*V. cucullata*) and **northern bog violet** (*V. nephrophylla*). Both of these species have green capsules and bear their inconspicuous summer flowers on erect stalks. However, the showy spring flowers of marsh blue violet are blue-violet with a darker centre, their lower petal is hairless, and their side petals have many club-shaped hairs with knobby tips. In northern bog violet, the spring flowers are deeper purplish (not darkened at the centre), their lower petal is silky-hairy, and their side petals have scattered, slender-tipped hairs. Marsh blue violet is found from Lake Superior to Newfoundland, south to Nebraska, Arkansas and Georgia, and the range of northern bog violet extends from British Columbia to Newfoundland, south to California, New Mexico, North Dakota, Iowa and New England.

V. nephrophylla

81

NORTHERN MARSH VIOLET
VIOLETTE PALUSTRE • *Viola palustris*

GENERAL: Hairless, perennial herb, **lacking leafy stems**; from long, **thread-like, horizontal rhizomes** and slender **runners** (stolons).

LEAVES: Alternate, arise directly **from rhizome**, hairless, **heart-shaped**, 2.5–6 cm wide, round-toothed.

FLOWERS: From rhizome, usually **taller than leaves**, **nodding**, with 5 unequal petals, of 2 types; showy spring flowers **pale lilac** (sometimes nearly white), **with darker veins**, 8–12 mm long, their lowest petal with a very short (2 mm long) rounded spur and their **side petals slightly hairy**; inconspicuous summer flowers, self-fertilizing (cleistogamous).

FRUITS: Capsules, 6–8 mm long, splitting open and projecting seeds.

WHERE FOUND: Wet or moist soil; from northern British Columbia to Newfoundland, south to California and Colorado in the west and to New Hampshire and Maine in the east.

V. palustris

NOTES: Lance-leaved violet (*V. lanceolata*) and northern white violet (*V. macloskeyi*, previously known as ***V. pallens***) have a similar growth form, with slender, horizontal rhizomes and thread-like runners, but their flowers are white, with dark purple veins. Lance-leaved violet has distinctive, lance-shaped leaves (unusual for a violet), that are 5–10 cm long in the spring, but reach 20–30 cm in length later in the summer. It grows in damp to wet, open sites and woods from Lake Superior to Nova Scotia, south to Nebraska, Texas and Florida. The leaf and flower stalks of northern white violet are often somewhat hairy and may be dotted with red. This species grows in wet to boggy sites from British Columbia and the southern Northwest Territories to Newfoundland, south to California, North Dakota, Alabama and South Carolina. • The genus name *Viola* means 'fragrant plants,' and it refers to the flowers, which are used to make perfumes and potpourri.

V. lanceolata

V. macloskeyi

KIDNEY-LEAVED VIOLET
VIOLETTE RÉNIFORME • *Viola renifolia*

GENERAL: Perennial herb, **lacking leafy stems**, **more or less hairy**; from long, **slender rhizomes** and **few**, short, thread-like **runners** (stolons).

LEAVES: Alternate, directly **from rhizome**, **hairy, kidney-shaped** (wider than long), rounded or abruptly pointed at tip.

FLOWERS: From rhizomes or runners, nodding, often **shorter than leaves**, with 5 unequal petals, of 2 types; showy spring flowers **white with brownish-purple lines**, **hairless**, their lowest petal with a spur (nectary); inconspicuous summer flowers self-fertilizing (cleistogamous), borne on short runners.

FRUITS: Ovoid, purplish capsules, splitting open and projecting brown seeds; stalks horizontal at first, erect at maturity.

WHERE FOUND: Swamps and cold woods; from southern Alaska to Newfoundland, south to Colorado, Wisconsin and Connecticut.

NOTES: Sweet white violet (*V. blanda*) is very similar to kidney-leaved violet, but its flowers are much taller than its leaves, and its leaves are heart-shaped (rather than kidney-shaped) and essentially hairless, with just a few minute, white hairs on the upper surface. Sweet white violet grows in moist rich woodlands, from Manitoba to southern Quebec and New Hampshire, south to Illinois and Georgia. • The presence of salicylic acid has made this plant a useful herbal medicine for treating heart attacks, strokes, skin infections and fungal infections, and it is useful as a pain killer. Ancient European herbalists and North American medicine men claim to have treated cancer with this plant.

V. blanda

V. renifolia

PURPLE LOOSESTRIFE
LYTHRUM SALICAIRE • *Lythrum salicaria*

GENERAL: Perennial herb, up to 150 cm tall; often **in large colonies**, from long, thick, woody taproots.

LEAVES: Opposite or whorled, 3–10 cm long, lance-shaped, **clasping stem somewhat**.

FLOWERS: Showy, purple, 7–12 mm wide, **5–6-parted**, in **elongated,** 10–40 cm long, **dense spikes**; mid-summer.

FRUITS: Capsules, numerous (as many as 900 on a healthy plant), containing approximately 100 seeds; late summer.

WHERE FOUND: Usually near cities, towns and roads, in ditches, wet meadows and marshes and on shores; from Newfoundland to southern British Columbia, south to Missouri and Virginia.

NOTES: Purple loosestrife is native to Europe and was introduced to the eastern coast of North America (via ship ballast unloaded on the shores) in the early 1800s. It outcompetes native wetland species, quickly colonizing large areas and often becoming the dominant species in wet meadows and shallow marshes. Research is currently being conducted into the use of 2 species of leaf-eating beetles and a species of root-eating weevil for biological control. Purple loosestrife is a very attractive plant and is sometimes sold at seed centers to gardeners and bee keepers, spreading this weed across the province. • Purple loosestrife is a noxious weed. It is rarely eaten by wildlife and it produces enormous amounts of viable, vigorously germinating seed.

WATER-WILLOW, WHORLED LOOSESTRIFE, SWAMP LOOSESTRIFE
DÉCODON VERTICILLÉ • *Decodon verticillatus*

GENERAL: Perennial, **aquatic**, **slightly shrubby herb; stems 4–6-sided**, 60–250 cm long, curved backward, **rooting at tips** in water or mud; submerged bark often thickened and spongy.

LEAVES: Opposite or whorled, **lance-shaped**, 5–13 cm long, 0.8–2.5 mm wide, short-stalked.

FLOWERS: Nearly 2.5 cm wide; 5 **pink-purple** petals; **sepals** with 5–7 pointed **teeth alternating** with 5–7 long, spreading **horns**; in several-flowered clusters (cymes) from leaf axils

FRUITS: Rounded, 3–5-celled capsules, about 5 mm in diameter.

WHERE FOUND: Usually at water's edge and drooping or trailing in shallow ponds, lakes, streams, marshes and sometimes swamps, often in very soft substrates; may form dense colonies or floating mats; from Minnesota and southern Ontario to Nova Scotia, south to Louisiana and Florida.

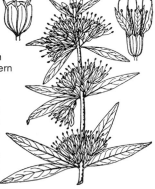

NOTES: Like many members of the loosestrife family, water-willow has styles and stamens of 3 lengths. This arrangement helps to ensure cross-pollination. • Water-willow seeds are eaten by waterfowl. • The genus name *Decodon* is derived from the Greek *deka*, meaning 'ten,' and *odous*, meaning 'tooth,' referring to the 10 teeth of the sepals. The species name *verticillatus* refers to the whorled (verticillate) leaves.

MEADOW BEAUTY, DEER-GRASS
Rhexia virginica

GENERAL: Perennial herb, 30–45 cm tall; **stems** stout, **4-angled**, often slightly **winged**; roots bear **tubers**.

LEAVES: Opposite, stalkless, **egg- to lance-shaped**, pointed at tip, rounded at base, 2–5 cm long, 1.2–2.5 cm wide, **sharp-toothed**.

FLOWERS: Showy, bright purple, with **8 conspicuous yellow stamens** protruding from centre; 4 rounded petals, 25–40 mm wide; calyx and stalk glandular-hairy; in few-flowered clusters (cymes) at stem tip and in upper leaf axils.

FRUITS: 4-celled capsules, splitting lengthwise into 4 parts; seeds numerous, coiled (**like tiny snail shells**).

WHERE FOUND: On open, damp, sandy, coarse gravelly or mucky shorelines; from central Ontario to Nova Scotia, south to Missouri, Alabama and Georgia.

NOTES: Meadow beauty may have migrated into Ontario from the Atlantic Coastal Plain during the retreat of the Wisconsin ice sheet. It survives here in a few open habitats where periodic flooding limits the growth of shoreline shrubs. The stabilization of water levels by dams and shoreline development threatens this species. • The leaves are said to have a sweetish and slightly acid taste, and the tubers a pleasant, nutty flavour, but this plant is uncommon in Ontario, and therefore it should not be collected.

MARSH WILLOWHERB
ÉPILOBE PALUSTRE • *Epilobium palustre*

GENERAL: Perennial herb, up to 80 cm tall; stems hairy at tip; from **long, slender stolons**.

LEAVES: Smooth, **untoothed,** 2–7 cm long, 2–15 mm wide; opposite at bottom of plant and alternate at tip.

FLOWERS: White or pink, 4–6 mm wide, **in upper leaf axils**, **5-parted**; **petals notched**; early to mid-summer.

FRUITS: Long, slender, hairy **capsules,** splitting open from tip to base; **seeds tipped with a tuft of long, white hairs**; mid-summer to early autumn.

WHERE FOUND: In marshes and swamps; from Newfoundland to Alaska, south to Oregon, Minnesota and Massachusetts.

NOTES: Two similar species of *Epilobium* are common in marshes and swamps in this region. **Northern willowherb** (*E. ciliatum*, inset photo) has toothed leaves, and it is found from Newfoundland to Alaska, south to California, Missouri and Maryland. **Narrow-leaved willowherb** (*E. leptophyllum*) has many small hairs on the upper surface of its leaves and on the upper parts of its stem. It grows from Newfoundland to Alaska, south to Oregon, Minnesota and Massachusetts • Deer and moose eat willowherbs. • The name *palustre* means 'of the marsh or swamp.'

85

BULBIFEROUS WATER HEMLOCK
CICUTAIRE BULBIFÈRE • *Cicuta bulbifera*

GENERAL: Delicate perennial herb, with slender, sparse branches, usually less than 50 cm tall, but up to 1 m; **stem base not strongly swollen**; vegetative **bulblets** clustered in upper leaf axils.

LEAVES: Alternate, **twice pinnately compound, with narrow segments up to 5 mm wide**.

FLOWERS: Numerous, small, white, in **flat-topped clusters (umbels)** up to 5 cm wide at tip of plant.

FRUITS: Flattened, small, broadly oval, with vertical ribs, rarely mature.

WHERE FOUND: In marshes, swamps and thickets and along lakeshores and streambanks; from Newfoundland to the Northwest Territories, south to Oregon, Ohio and Virginia.

NOTES: Water-hemlock is probably one of the most **violently poisonous** plants in temperate North America. Even the juice from crushed plants can cause poisoning. Symptoms can occur in minutes and include abdominal pain, violent convulsions, fever, paralysis, respiratory failure and death. Most reported water hemlock poisonings have resulted from eating a closely related species, **spotted water hemlock** (*C. maculata*, p. 87), which grows in the same habitat, but is larger, lacks bulblets, and has many well-developed flat-topped clusters (umbels) and wider leaflets. The same toxins are present in bulbiferous water hemlock. Avoid handling this plant and always wash your hands after contact. • The vegetative bulblets are the main means of reproduction, and they are sometimes eaten by moose. Waterfowl eat the small seeds. • In Ojibway, bulbiferous water hemlock is called *musquash root*. It was used medicinally and smoked before going deer hunting to attract bucks. Early records say of *Cicuta* that 'when crushed and placed on a man's genitals, it does away for the lust for women.' • The species name *bulbifera* refers to the vegetative bulblets, which are a good identification feature.

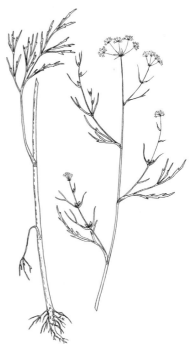

SPOTTED WATER HEMLOCK
CICUTAIRE MACULÉE • *Cicuta maculata*

GENERAL: Stout perennial herb, up to 2 m tall; **stems with purple blotches**, taller than leaves, **thickened, hollow and chambered at base**; from thick, tuberous roots.

LEAVES: Alternate, **2–3 times pinnately compound**; leaflets narrow, 3–10 cm long, 0.5–3.5 cm wide, toothed.

FLOWERS: Numerous, small, white; in **flat-topped clusters (umbels)** 5–10 cm wide.

FRUITS: Flattened, broadly oval, 2–4 mm wide, with vertical ribs, pale brown.

WHERE FOUND: In marshes, ditches, thickets and swamps and on shores; Nova Scotia to Alaska, south to Texas, Tennessee and North Carolina.

NOTES: Spotted water-hemlock is one of the most **violently poisonous** plants in North America. The juice from crushed plants can cause poisoning if transferred from hands to mouth. The symptoms can occur in minutes and include abdominal pain, violent convulsions, fever, paralysis, respiratory failure and death. Most poisonings have resulted from eating spotted water hemlock, but the same toxins are present in bulbiferous water hemlock (*C. bulbifera*, p. 86). **Avoid handling these plants and always wash your hands after contact.**

WATER PARSNIP
BERLE DOUCE • *Sium suave*

GENERAL: Perennial herb, up to 2 m tall; stems solitary, corrugated; from fibrous roots.

LEAVES: Alternate, compound, **once-divided**, with 7–17 toothed, 5–10 cm long leaflets; **finely-divided, submerged leaves sometimes present; leaf stalks sheathing.**

FLOWERS: Very small, 3–4 mm wide, white, 5-parted, in **flat-topped, 3–12 cm wide clusters (umbels) at stem tips**; mid- to late summer.

FRUITS: Oval, 2–3 mm long, with corky ribs; late summer.

WHERE FOUND: In marshes, swamps and ditches and along lakeshores, rivers and streams, often in standing water; Newfoundland to Alaska, south to California, Ohio and Florida.

NOTES: Water parsnip is **very similar to the water hemlocks (*Cicuta* spp.,** above and p. 86) **which are extremely poisonous**, but water hemlocks have twice-divided (rather than once-divided) leaves. Because of this close similarity, **extreme care** should be taken when handling these plants. • Water parsnip is said to be edible, but this is questionable, and it should not be eaten because of the possibility of confusing it with water hemlock. • The Ojibway smoked water parsnip seeds over the fire to drive away and blind *Sokenau*, the evil spirit that brings bad luck to hunters.

87

Dogwood Family (Cornaceae)

BUNCHBERRY
CORNOUILLER DU CANADA • *Cornus canadensis*

GENERAL: Erect, 10–20 cm tall, perennial herb; spreading from underground rhizomes, often forming large colonies.

LEAVES: Elliptic or egg-shaped, 4–8 cm long, not toothed, tapering to point at both ends, opposite, but appearing as a **whorl of 4–6 leaves at tip of stem**; **side veins conspicuous and parallel to leaf edge**; 1 or 2 pairs of smaller, scale-like leaves often on stem below.

FLOWERS: Small, greenish to purplish, 4-petalled, in solitary clusters at stem tips resembling a single flower with **4 showy, 1–2 cm long, white, petal-like bracts surrounding a tight cluster of tiny flowers**; early to mid-summer.

FRUITS: Bright red, pulpy, berry-like drupes in a cluster at the stem tip; mid- to late summer.

WHERE FOUND: In upland forests and hardwood and conifer swamps; from Newfoundland to Alaska, south to California, South Dakota, Ohio and Pennsylvania.

NOTES: The fruits are an important food for some song-birds, grouse, rock voles and chipmunks. • Although rather bland and somewhat mealy, bunchberry fruits are edible and can be used in puddings. • Bunchberry has been used to treat colds, and a tea made from its roots has been used as a remedy for colic in infants. • It is easy to overlook the true flowers of bunchberry. The striking white 'petals' are really large bracts surrounding a cluster of tiny flowers. The showy white bracts make the flowers more noticeable to passing insects and lure them to the plant. In autumn, the flowers mature into a cluster of scarlet berry-like drupes—hence the name 'bunchberry.'

PINK PYROLA, BOG WINTERGREEN
PYROLE À FEUILLES D'ASARET • *Pyrola asarifolia*

GENERAL: Low, evergreen, perennial herb; from branched rhizomes.

LEAVES: Basal, leathery, shiny, 3–6 cm wide, rounded to **kidney-shaped; stalks longer than blades**.

FLOWERS: Pink, nodding, short-stalked, 3 cm wide, with 5 sepals and 5 petals; loosely arranged in an elongating cluster (raceme) on a long, leafless stalk 15–30 cm tall; early to mid-summer.

FRUITS: Nodding capsules, with persistent styles; mid-summer.

WHERE FOUND: In moist upland forests and swamps and along forested streams and rivers; from Newfoundland to Alaska, south to Oregon, South Dakota, and New England.

NOTES: Other pyrolas are typically found in drier, upland forests. **One-sided wintergreen** (*Orthilia secunda*) has greenish white flowers arranged on 1 side of the stalk. It grows in moist to dry forests, from Newfoundland to Alaska, south to Mexico, Ohio and Virginia. • Pink pyrola fruits are eaten by grouse. • Native people used pink pyrola to make a tea and to treat colds, wounds, weak nerves, mouth sores and many other ailments. • The genus name *Pyrola* means 'leaves like those of the pear tree,' and the species name *asarifolia* means 'leaves like those of wild ginger.'

STARFLOWER
TRIENTALE BORÉALE • *Trientalis borealis*

GENERAL: Low, perennial herb, 10–20 cm tall, with a **whorl of 5–10 deciduous leaves at tip of a single, erect, unbranched stem**; from slender rhizomes.

LEAVES: Narrow, somewhat **lance-shaped**, long-pointed at tip and narrowed at base, 4–10 cm long, shiny, hairless, not toothed; usually with 1 small, scale-like leaf about half-way up stem.

FLOWERS: Star-shaped, white, up to 1.5 cm across, usually with 7 delicate petals, 1–3 flowers on slender stalks from centre of leaf whorl; late spring to early summer.

FRUITS: 5-segmented capsules, with many tiny, black seeds; mid-summer.

WHERE FOUND: In upland forests and richer conifer swamps; from Newfoundland to the Yukon Territory, south to California, Minnesota and Virginia.

NOTES: Starflower has been used to make a medicinal tea to treat tuberculosis.

TUFTED LOOSESTRIFE
LYSIMAQUE THYRSIFLORE • *Lysimachia thyrsiflora*

GENERAL: Perennial herb; stems unbranched, 20–60 cm tall; from long rhizomes.

LEAVES: Opposite, lance-shaped, **black spotted**, 5–12 cm long.

FLOWERS: Yellow, sometimes streaked with black or purple, 5-parted, 6 mm wide, in short, dense **spikes on 2–4 cm stalks from leaf axils;** early to mid-summer.

FRUITS: Oval capsules, splitting from tip; seeds angular, few.

WHERE FOUND: In marshes, wet fields and beaver meadows and on lakeshores; from Nova Scotia to Alaska, south to California, Colorado and Virginia.

NOTES: **Fringed loosestrife** (*L. ciliata*, inset photo) has a fringe of hairs on its leaf edges and stalks. Also, its leaves are not spotted with glands and its flowers are solitary on long stalks from leaf axils. Fringed loosestrife grows on rich shores and in swamps, from Nova Scotia to southern British Columbia, south to Oregon, Texas and Florida. • **Purple loosestrife** (*Lythrum salicaria*, p. 84) is not related to *Lysimachia* species • The common name means 'ending strife' and the genus was named after King Lysimachos of Thrace (360–281 BC) who pacified a wild bull with a piece of loosestrife.

SWAMP CANDLES
LYSIMAQUE TERRESTRE • *Lysimachia terrestris*

GENERAL: Perennial herb, 40–80 cm tall; stems square; bulblets developing in leaf axils; from long, stoloniferous rhizomes.

LEAVES: Opposite, narrowly lance-shaped, 5–10 cm long, **black-dotted**.

FLOWERS: Yellow, **marked with dark lines**, 5–6-parted, 1–2 cm wide, on 8–15 mm long stalks; erect, elongated clusters (racemes) 10–30 cm long, at the stem tip; mid- to late summer.

FRUITS: Capsules, containing several small seeds.

WHERE FOUND: In floating mat fens and on shores; from Newfoundland to southern Manitoba, south to Iowa, Kentucky and Georgia.

NOTES: A very distinctive flower that rises above the surrounding sedges.

BUCKBEAN, BOGBEAN
MÉNYANTHE TRIFOLIÉ • *Menyanthes trifoliata*

GENERAL: Perennial herb, 5–30 cm tall; **stems trailing**; from thick rhizomes covered with old leaf bases.

LEAVES: Basal, 3–6 cm wide, compound, with 3 elliptic leaflets, with wavy edges; **stalks long, with sheathing base**.

FLOWERS: White, with 5 sepals and **5 fringed petals,** on 5–20 mm long stalks; in elongated clusters (racemes) on leafless stalks, taller than leaves; spring to early summer.

FRUITS: Thick, corky-walled capsules, 6–10 mm long, filled with many yellowish-brown seeds; mid- to late summer.

WHERE FOUND: Typically in fen pools and hollows or at outer edge of floating mats, occasionally as an emergent in shallow marshes; from Newfoundland to Alaska, south to California, Missouri and Delaware.

NOTES: Buckbean is very bitter to eat, but it has been used to make a tonic. The rhizomes can be boiled, dried and pounded into a nutritious but bitter flour (heat does not destroy the bitterness). Medicinally, it has been used to treat rheumatism, liver, stomach and gall bladder problems (it increases the flow of gastric secretions) and fevers. • The genus name *Menyanthes* means 'moon flower' and the species name *trifoliata* means 'with 3 leaflets.'

SWAMP MILKWEED
ASCLÉPIADE INCARNATE • *Asclepias incarnata*

GENERAL: Perennial herb, up to 1.5 m tall, with **milky sap**; from spreading rhizomes.

LEAVES: Opposite, entire, prominently veined, 7.5–15 cm long, 1–4 cm wide.

FLOWERS: Rose or purple, in flat-topped clusters (umbels) at stem tips; **5 petals and 5 sepals bent backward; erect crown** in centre of flower consists of 5 horned hoods, with **horns longer than hoods**; mid-summer.

FRUITS: Large pods, 5–9 cm long, produced by only a few flowers; many **seeds, with tufts of long, fluffy, white hairs**, dispersed by wind in late summer.

WHERE FOUND: In swamps, marshes, streambanks, ditches and fens; from Nova Scotia to southern Manitoba, south to New Mexico, Tennessee and South Carolina.

NOTES: Common milkweed (*A. syriaca*) has broader leaves (5–11.5 cm wide), pale pink flowers with horns shorter than their hoods and larger (7–13 cm long) pods covered with woolly hairs and short, soft 'spines.' It grows under drier conditions in fields and on roadsides, from southern Manitoba to Nova Scotia, south to Kansas and Georgia. • The seeds are eaten in small amounts by ducks and other waterfowl, and their fluffy hairs are used by birds to line their nests. Milkweeds produce copious amounts of nectar that attracts many insects. Bees are the chief pollinators, feeding on the nectar and filling the pollen sacs on their legs. • The milky sap contains a glycoside that is **poisonous** to many animals. Several insects, such as milkweed bugs, milkweed beetles and monarch butterfly caterpillars, feed on the leaves and acquire protection from predators by retaining these poisons. • Native people used the stem fibres to make twine, and a tea made from the roots was said to drive worms from the body within an hour. • The species name *incarnata* means 'flesh-coloured,' and it refers to the flowers.

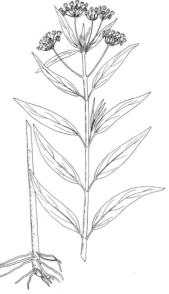

A. syriaca

A. incarnata

SQUARE-STEMMED MONKEY FLOWER
MIMULE À FLEURS ENTROUVERTES • *Mimulus ringens*

GENERAL: Hairless, perennial herb; stems erect, **4-sided, often winged** on angles, 30–90 cm tall; from rhizomes.

LEAVES: Opposite, sharply toothed, oblong to lance-shaped, 5–10 cm long, 1–2 cm wide, pointed at tip, stalkless, **often heart-shaped and clasping** stem at base.

FLOWERS: Violet (rarely white), **tubular, 2-lipped**, 2–4 cm long; upper lip erect or bent back, 2-lobed; lower lip 3-lobed, short-hairy within; **stalks slender, longer** than flowers; July.

FRUITS: Oblong capsules enclosed by persistent sepals; seeds numerous, tiny, oblong, with a network of raised veins.

WHERE FOUND: In rich muck of marshes, beaver ponds, swamps, lakeshores, riverbanks, ditches and fens; from eastern Saskatchewan to Nova Scotia, south to Texas and Georgia.

NOTES: Monkey flowers could be confused with members of the mint family because the stems are square, the leaves are opposite, and the flowers are prominently 2-lipped. The flowers produce capsules however, rather than 4 1-seeded nutlets. • **Small-flowered agalinis** (*Agalinis paupercula*, formerly known as *Gerardia paupercula*) is also in the fig-wort family, but it has smaller, purple-pink flowers with 5 regular petals. It is found on sandy shores and in thickets from southern and central Ontario to New Brunswick, south to Florida. There are 4 other species of *Agalinis* in Ontario, found mainly in the south. • Monkey flowers are pollinated by bumble-bees. The stigma is very sensitive, and when it is touched, the 2 lips of the flower close to help ensure cross-pollination. • Monkey flower gets its common name from the flower, which resembles the face of a grinning monkey. The genus name *Mimulus* means 'little mimic, actor, or buffoon,' in reference to the 'grinning' flowers. The species name *ringens* means 'gaping,' in reference to the mouth of the flower (or the monkey).

M. ringens

Agalinis paupercula

Agalinis paupercula

TURTLEHEAD
GALANE GLABRE • *Chelone glabra*

GENERAL: Perennial herb; stems erect, slender, **4-sided**, 30–90 cm tall; branches erect.

LEAVES: Opposite, sharply toothed, short-stalked, narrowly to broadly **lance-shaped**, slender-pointed at tip, tapered to base, 7–20 cm long, 1–4 cm wide.

FLOWERS: White or pinkish, **tubular, 2-lipped**, about 2.5 cm long; upper lip concave; lower lip 3-lobed, hairy within; in **dense, hairless spikes** at stem tip and from leaf axils.

FRUITS: Ovoid capsules, about 1.2 cm long, with many, flattened, winged seeds.

WHERE FOUND: In swamps, wet woods, thickets, ditches, streambanks and lakeshores; from southeastern Manitoba to Newfoundland, south to Missouri and Georgia.

NOTES: Turtlehead could be mistaken for a member of the mint family when it is not in flower, but it has rounded, not square, stems. • Caterpillars of the Baltimore butterfly feed on the leaves of turtlehead. • The genus name *Chelone* was derived from the Greek *chelone*, 'a turtle,' because the flowers were thought to resemble the heads of a tiny turtles. The species name *glabra* refers to the hairless (glabrous) bracts and calyxes of the flower clusters.

NORTHERN BUGLEWEED
LYCOPE À UNE FLEUR • *Lycopus uniflorus*

GENERAL: Perennial herb, may have a slight minty fragrance; **stems square**, usually unbranched; from rhizomes with short, thick tubers.

LEAVES: Opposite, short-stalked, up to 6 cm long, with widely spaced, **coarse teeth**; pairs at right angles to each other.

FLOWERS: Small, irregular, white, **in axils of mid to upper leaves**; early to mid-summer.

FRUITS: Nutlets, with corky ridge on edges, longer than lobes of sepals; mid- to late summer.

WHERE FOUND: On shores and in marshes, wet fields, swamps, ditches and fens; from Newfoundland to British Columbia, south to California, Oklahoma and North Carolina.

NOTES: Water horehound (*L. americanus*) has leaves that are deeply lobed (often the lowest teeth are the largest), short-stalked and up to 8 cm long. Its rhizomes lack tubers and the lobes of its sepals are longer than the nutlets. Water horehound grows in low, wet areas from Newfoundland to British Columbia, south to California, Texas and Florida. • The tubers of *Lycopus* are an important food for muskrats. • The tubers have been used like cucumbers in salads and preserved like pickles. Native people called them 'crow potatoes.' • The name *Lycopus* is taken from the Greek for 'wolf's foot,' and it refers to the lobed lip on the flower, which was thought to be similar to a wolf foot or foot print.

L. americanus L. uniflorus

WILD MINT
MENTHE DES CHAMPS • *Mentha arvensis*

GENERAL: Perennial herb, with **strong mint fragrance**; stems square, slightly to densely hairy, 20–80 cm tall; from rhizomes.

LEAVES: Opposite, toothed, oval, 2–8 cm long, 1–4 cm wide, short-stalked; **pairs at right angles to each other.**

FLOWERS: Small, irregular, white, pinkish or lilac, in whorls in axils of leaves; mid-summer.

FRUITS: Very small nutlets, hidden by sepals; late summer to early autumn.

WHERE FOUND: In wet fields and swamps and on shores; from Newfoundland to Alaska, south to California, Virginia and Delaware.

NOTES: Grouse, green-winged teals and muskrats eat the nutlets. • Mint leaves are used to make tea and to season foods. • The active ingredient in wild mint is menthol, which has been used for soothing the stomach, aiding digestion and relieving intestinal gas. Native people used this plant to treat colds, coughs and fevers, as a blood remedy and to treat pneumonia. Mint plants can be hung upside down to dry and then stored for long periods.

COMMON SKULLCAP
SCUTELLAIRE À FEUILLES D'ÉPILOBE • *Scutellaria galericulata*

GENERAL: Perennial, **non-aromatic** herb; **stems square**, 20–80 cm tall; from rhizomes.

LEAVES: Opposite, 2–6 cm long, 1–2 cm wide, lance-shaped, short-stalked, hairy beneath, slightly toothed.

FLOWERS: Tubular, 1–2 cm long, irregular, with an **arching, hooded upper lip and flaring lower lip, blue with white markings, paired in leaf axils**; early to mid-summer.

FRUITS: Oval, golden-brown nutlets; mid- to late summer.

WHERE FOUND: On shores and fen edges and in marshes, wet fields, beaver meadows and ditches; from Newfoundland to Alaska, south to California, New Mexico and Delaware.

NOTES: This species is also known as *S. epilobiifolia*. • **Mad dog skull-cap** (*S. lateriflora*) grows in low, wet areas from Newfoundland to British Columbia, south to California and Georgia, but it is less common and is usually found along rocky shorelines. It is distinguished by its flowers, which are on long branches from the leaf axils. • Common skullcap has been used by native people and Europeans to treat spasms, restlessness, insomnia, excitability and heart problems. • The name 'skullcap' refers to the hump on the upper lobe of the calyx. The blue flowers have white markings that sometimes resemble skulls. The common name, 'mad dog skullcap,' comes from the historical use of this plant in treating people with rabies.

FRAGRANT BEDSTRAW
GAILLET À TROIS FLEURS • *Galium triflorum*

GENERAL: Perennial herb, sweet-smelling when dried; **stems 4-angled, trailing or ascending**, branching, 30–100 cm long, slightly roughened on angles; from slender rhizomes.

LEAVES: In whorls of 6, narrowly **oval to lance-shaped**, 1-nerved, **bristle-tipped**, 2–8 cm long, 4–12 mm wide.

FLOWERS: Small, **greenish**; petals fused at base, with 4 spreading lobes; in **widely branching clusters** (cymes) from leaf axils and stem tips; June–July.

FRUITS: Pairs of small, rounded, 1-seeded nutlets, **rough-hairy** with many hooked bristles, 3–4 mm long; July–August.

WHERE FOUND: In rich swamps, thickets and upland forests; from Alaska to Newfoundland, south to Mexico and Florida.

NOTES: **Rough bedstraw** (*G. asprellum*) also has rough stems and bristle-tipped leaves mostly in whorls of 6, but it is much more sprawling and bristly, with long, hooked bristles on its stems and almost prickly edges on its smaller (8–16 mm long) leaves. Its nutlets are hairless. Rough bedstraw grows in swamps, thickets and ditches from Ontario to Newfoundland, south to Nebraska and North Carolina. • These plants are fragrant when dry, and they have been mixed with mattress straw, hence the common name 'bedstraw.' Their vanilla-like fragrance comes from coumarin compounds, which help to prevent blood from clotting. • A Eurasian species, **yellow bedstraw** (*G. verum*), was believed to be the plant that filled the manger in Bethlehem. • Bedstraws belong to the same family as coffee, and the seeds of **common bedstraw** (*G. aparine*) make the best coffee substitute of any plant in Canada. Other species could probably be used as well, but it takes considerable work to gather the tiny seeds. Bedstraws have been used as potherbs, and the stems and leaves of species with curved bristles were used in Europe to strain straw, hair and other foreign material from fresh milk. • The name *Galium* comes from the Greek *gala*, 'milk,' which may refer to the traditional use of yellow bedstraw with rennet to curdle milk for cheese-making.

G. triflorum

G. asprellum

SMALL BEDSTRAW
GAILLET TRIFIDE • *Galium trifidum*

GENERAL: Low, sprawling, perennial herb, with whorled leaves; **stems weak**, often branched, 20–60 cm long, climb on other plants and cling to clothing with tiny hooks.

LEAVES: Mostly in **whorls of 4, 1-veined, narrow, blunt-tipped**, 5–20 mm long.

FLOWERS: 1 mm wide, with **3 white petals, stalked**, in clusters of **1–3 at stem tips or in leaf axils**; throughout summer.

FRUITS: Black, 1–2 mm thick, 1-seeded; throughout summer.

WHERE FOUND: In fens and swamps and on shores; from Newfoundland to Alaska, south to California, Texas and Georgia.

NOTES: The other wetland species of *Galium* all have whorled leaves and 3- or 4-parted flowers. • **Labrador bedstraw** (*G. labradoricum*) has 4-parted flowers and reflexed leaves. It is found from Newfoundland to the Northwest Territories, south to Minnesota, Ohio and New Jersey. • **Marsh bedstraw** (*G. palustre*) has 4-parted flowers in clusters of more than 5. It grows in wet meadows and thickets and on shores from Newfoundland to northern Ontario, south to Wisconsin, Pennsylvania, and New England. • Ducks, grouse and muskrats occasionally eat small bedstraw fruits. • Native people used the roots of some bedstraws to make a beautiful red dye, and they used the plants to make a pungent medicinal tea (with a persistent taste), to treat eczema, ringworm and respiratory illnesses. This is also a popular herbal remedy in Europe, used to treat skin disorders, urinary obstructions, kidney problems and colds. • The genus name *Galium* means 'milk,' and it refers to the flowers of a European species, wild madder (*G. mollugo*), which was traditionally used to curdle milk for making cheese. Wild madder grows in southern Ontario as a weed.

MARSH BELL FLOWER
CAMPANULE FAUX-GAILLET • *Campanula aparinoides*

GENERAL: Slender, perennial herb; **stems reclining or weak, 3-angled**; from long, thin rhizomes.

LEAVES: Alternate, narrow, rough or smooth, gradually shorter and narrower toward tip of plant, sparsely toothed or toothless.

FLOWERS: White or pale blue, bell-shaped, 1 cm long, **5-parted**, with 5 petal lobes and 5 triangular sepals, solitary, on long stalks; early to mid-summer.

FRUITS: Capsules, split open at bottom to release many small seeds; mid- to late summer.

WHERE FOUND: In marshes and wet meadows; from Nova Scotia to Saskatchewan, south to Colorado, Iowa and Georgia.

NOTES: Common bell flower (*C. rotundifolia*) differs from marsh bell flower by having larger, dark blue flowers and erect, cylindrical stems. It grows in a wide range of habitats, including upland forests, sand beaches and rocky shores, from Alaska to Newfoundland, south to northern California, northern Mexico, Nebraska and New Jersey. • Like **bedstraws** (*Galium* **spp.**, above), marsh bell flower has tiny hooks that cause its stems to cling to clothing. • The leaves are occasionally eaten by deer. • The genus name *Campanula* means 'little bell,' and it refers to the flower.

SLENDER WHITE ASTER, RUSH ASTER
ASTER BORÉALE • *Aster borealis*

GENERAL: Delicate, perennial herb; stems slender, usually unbranched, 10–50 cm tall in northern Ontario and up to 1 m tall in southern Ontario; from very thin rhizomes.

LEAVES: Alternate, **4–11 cm long, 1 cm wide, smooth, hairless, curved upward at edges**; as with most asters, lower leaves die and fall off during flowering.

FLOWERHEADS: Several, 2.5 cm wide, white, blue or lavender, with yellow or purple centres, on upper ⅓ of plant; stalks often hairy; late summer.

FRUITS: Achenes, tipped with a tuft of long, white, hair-like bristles (pappus).

WHERE FOUND: In cold fens, swamps, marshes and lakeshores; from Nova Scotia to Alaska, south to Idaho, Minnesota and New Jersey.

NOTES: This species was formerly known as *A. junciformis*. • **Lance-leaved aster** (*A. lanceolatus*, p. 99) grows in similar habitats, and it is distinguished by its thick stems and its rhizomes and many clustered flowerheads. • **Flat-topped white aster** (*A. umbellatus,* inset photo) is a tall, white-flowered aster with flowerheads in a spreading, flat-topped cluster. It is distinguished from similar asters by the hair-like bristles (pappus) on its achenes, which are borne in 2 distinct rows of hairs (a long inner row and a short outer row), rather than in 1 row. Flat-topped white aster grows on shores and in beaver meadows and other wet areas, from Alberta to Newfoundland, south to Nebraska, Kentucky and Georgia.

PURPLE-STEMMED ASTER
ASTER PONCEAU • *Aster puniceus*

GENERAL: Large, branched, perennial herb, **covered in dense hairs; stems reddish to purple**, 40–170 cm tall; from thick, short rhizomes.

LEAVES: Smooth or hairy, 7.5–15 cm long, 1.3–3.8 cm wide, **clasping stem**, toothed or toothless.

FLOWERHEADS: Pale blue to purple, with yellow central discs which turn purple, up to 3.5 cm wide; size and number of flowerheads varies.

FRUITS: Achenes, tipped with tufts of long, white, hair-like bristles (pappus).

WHERE FOUND: In swamps, ditches and other wetlands; from Newfoundland to Alberta, south to North Dakota, Alabama and Georgia.

NOTES: The variation in stem height and flower size and number is related to growing conditions. • Purple-stemmed aster, with its branched, purple stem and large, clasping leaves, is not likely to be confused with other asters. • The large amount of nectar and pollen produced by asters attracts many bees, flies, beetles and other insects. Ambush bugs and spiders lie waiting in these plants for pollen-gathering insects. • Traditionally, asters have been associated with good fortune and have been used to keep evil spirits away and to help heal wounds. • The species name *puniceus* means 'crimson-coloured,' and it refers to the stem colour (actually more purple in our plants).

LANCE-LEAVED ASTER
Aster lanceolatus

GENERAL: Perennial herb, 60–150 cm tall, with **many leaves and flowers**; **stems and rhizomes thick**.

LEAVES: Lance-shaped, up to 14 cm long; lower leaves fall off at flowering time.

FLOWERHEADS: Numerous, up to 2.5 cm wide, white to pale blue, with pale yellow centres; bristles of fruits (pappus) extend above central disc, making it look bristly and white or pale yellow.

FRUITS: Achenes, tipped with a tuft of long, white, hair-like bristles (pappus).

WHERE FOUND: In marshes, fens and swamps, as well as in open fields and roadside ditches; from Newfoundland to Saskatchewan, south to Kansas and North Carolina.

NOTES: Lance-leaved aster is one of the most difficult asters to identify because of the great variability in its form: the plant can be covered in hair or it can be hairless; it can be tall or short; it can have different flower colours; and it can have long or short leaves. All of these variations are influenced by the growing conditions. • The species name *lanceolatus* refers to the lance-shaped leaves.

BOG ASTER
ASTER DES BOIS • *Aster nemoralis*

GENERAL: Perennial herb, 10–60 cm tall, **few-flowered**; stems solitary, hairy, rarely branched; from slender rhizomes.

LEAVES: Alternate, **1–7 cm long, linear, 5 mm wide, hairy, crowded** on stem; **edges rolled under**.

FLOWERHEADS: Usually **solitary on long stalks**, up to 4 cm wide, pale pink to violet, with yellow or purple central discs.

FRUITS: Achenes, tipped with a tuft of long, white, hair-like bristles (pappus).

WHERE FOUND: In poor fens and on shores, often intermixed with ericaceous shrubs and sedges on floating mats; from Newfoundland to Lake Superior, south to Michigan and New Jersey.

NOTES: The flowers of asters dry very well and are often used in dried flower arrangements. • The species name *nemoralis* means 'of the woods,' and it refers to the conifer swamps that this aster inhabits in parts of its range.

SWAMP THISTLE
CHARDON MUTIQUE • *Cirsium muticum*

GENERAL: Stout, biennial herb, up to 2 m tall; **stems not prickly**; from stout roots.

LEAVES: Alternate, 20 cm long, 5.5 cm wide, often deeply cut into lance-shaped lobes, **weakly spiny,** may be slightly cobwebby beneath.

FLOWERHEADS: Purple or pink, about 3–4 cm high and wide, numerous, with **fine, white, cobwebby hairs on bracts at base**; late summer.

FRUITS: Dry achenes, 5 mm long, with a tuft of white, hair-like bristles (pappus) at tip.

WHERE FOUND: In swamps, wet meadows, moist woods and thickets; Newfoundland to Saskatchewan, south to Tennessee and North Carolina.

NOTES: Swamp thistle is distinguished from other northern Ontario thistles by the combination of non-prickly stems and cobwebby hairs at the base of its flowerheads.

NORTHERN SWEET COLTSFOOT
PÉTASITE PALMÉ • *Petasites frigidus*

GENERAL: Perennial herb; often forming patches from spreading by rhizomes.

LEAVES: Basal, palmately veined, 5–40 cm wide, long-stalked, deeply cut into 5–7 segments, white woolly on lower surface.

FLOWERHEADS: 8–13 mm wide, with many cream-coloured ray florets, sometimes tinged pink; bracts in 1 row, green, parallel-veined; in **elongating cluster** on a **stout, scaly stalk 15–20 cm tall; early spring, before leaves.**

FRUITS: Linear achenes, 5–10 ribbed, with a tuft of **white hair-like bristles (pappus) at tip, in cottonball-like heads**; spring.

WHERE FOUND: In swamps, moist woods and wet meadows; from Newfoundland to Alaska, south to California, Minnesota and Massachusetts.

NOTES: This species includes *P. palmatus.*
• **Arrow-leaved coltsfoot (*P. sagittatus*, inset photo)** has distinctive, arrow-head-shaped leaves that are shallowly toothed along their edges, rather than deeply cut into lobes. It grows in wet, lime-rich areas, such as wet meadows, cedar swamps and low woods, from Alaska to Labrador, south to Washington, Colorado, Wisconsin and central Quebec. • Northern sweet coltsfoot is one of the earliest plants to flower in the spring. • The genus name *Petasites* means 'sun hat,' and it refers to the large basal leaves. The species name *frigidus* means 'cold,' and it refers to the cold, moist habitat of this plant.

SPOTTED JOE-PYE WEED
EUPATOIRE MACULÉE • *Eupatorium maculatum*

GENERAL: Robust, perennial herb, 60–200 cm tall; stems spotted with purple or solid purple.

LEAVES: Lance-shaped, 6–20 cm long, 2–9 cm wide, with **1 central main vein,** toothed, **in whorls of 3–7.**

FLOWERHEADS: Purple to pink, in flat-topped, 15–20 cm wide clusters of 9–22 flowerheads; only **tubular (disc) florets** (no ray florets); late summer.

FRUITS: Glandular-dotted achenes, 3–4.5 mm long, with a tuft of white, hair-like bristles (pappus) at tip.

WHERE FOUND: In wet ditches, moist old fields and swamps and on riverbanks; from Newfoundland to southern British Columbia, south to Utah, Iowa and North Carolina.

NOTES: In the winter this plant can be easily identified by the tiny white knobs at the tips of its branches (where its seeds were attached) and by the leaf scars on its stem (where its leaves were attached). • **Boneset (*E. perfoliatum*)** resembles spotted Joe-Pye weed, but it is easily distinguished by its dull white flowers and opposite, lance-shaped leaves, which join at their bases to encircle the stem. It grows in swamps, riverbanks, moist thickets and ditches from southeastern Manitoba to Nova Scotia, south to Texas and Florida. • Moth larvae live within the leaves, creating brown or black patches where they have eaten the inner leaf cells. • The species name *maculatum* means 'spotted,' and it refers to the purple-spotted stem. The common name comes from an American native man named 'Joe Pye' who used this plant to cure the colonists of typhus fever.

E. perfoliatum

E. maculatum

NORTHERN BOG GOLDENROD
VERGE D'OR DES MARAIS • *Solidago uliginosa*

GENERAL: Perennial herb, 30–120 cm tall; stems hairless below flower cluster; from thick, branching rhizomes.

LEAVES: Alternate; **lower leaves lance-shaped, 6–35 cm long, 6–60 mm wide, with a long, sheathing** stalk; upper leaves smaller and stalkless.

FLOWERHEADS: Yellow, on loosely ascending branches, highly variable in size and arrangement.

FRUITS: Hairless achenes, with a tuft of white, hair-like bristles (pappus) at tip.

WHERE FOUND: Usually in fens, swamps and wet depressions in bedrock; from Newfoundland to southern Manitoba, south to Wisconsin, Ohio and New Hampshire.

NOTES: Grass-leaved goldenrod (*Euthamia graminifolia*, also known as *S. graminifolia*) is recognized by its many 3–5-ribbed, linear-lance-shaped leaves (3–12 mm wide), and its flat-topped clusters (corymbs) of small, stalkless, yellow flowerheads. It grows on shores and in fields and ditches from British Columbia and the southern Northwest Territories to Newfoundland, south to New Mexico, Missouri and North Carolina. • **Canada goldenrod** (*S. canadensis*) has spreading flower clusters (panicles) with arching branches, and its lance-shaped leaves are larger near the middle of the stem than at the base. The stems are hairless or have a few short hairs near the top. The leaves are neither clasping, nor triple-nerved, and there are no basal leaves. Canada goldenrod grows in wet pastures, open woods, thickets, roadsides and on streambanks from Alaska to Newfoundland, south to California, Texas and Florida. • **Rough goldenrod** (*S. rugosa*) is very similar to Canada goldenrod, but its stems are silky-hairy with long, brownish hairs. Rough goldenrod gets its name from its leaves, which are conspicuously wrinkled (rugose) beneath. It grows in fields, open woods and damp thickets and along roadsides from Ontario to Newfoundland, south to Texas and Florida. • Goldenrods have an undeserved reputation for causing allergy problems. The pollen of goldenrod is large and sticky, and it is dispersed by insects, not the wind. Ragweed and other weeds are the source of allergy discomfort, but the colourful yellow fields of goldenrod get the blame. • Insects often lay eggs in goldenrod plants, producing galls. A midge causes black spots or blister galls on the leaves, moths cause the elliptic galls on the stems, and small flies cause the rounded galls on the stem. The galls attract many other insects that live in the galls, including inquilines (living with the gall-maker) and predators. • A fascinating relationship exists between leafhoppers and ants. The leafhoppers supply the ants with their excess sap and the ants protect the leafhoppers'

food (the goldenrod leaves) from other insects, such as the goldenrod beetle. • The genus name *Solidago* means 'to make whole,' and it refers to the historical medicinal uses of some species. The species name *uliginosa* means 'of swamps.'

S. canadensis

Euthamia graminifolia

NODDING BUR-MARIGOLD
NODDING BEGGAR'S TICKS
BIDENT PENCHÉ • *Bidens cernua*

GENERAL: Leafy, robust, annual herb; stems hairless or hairy, 10–100 cm tall.

LEAVES: Lance-shaped, 5–15 cm long, **opposite**, stalkless, often basal, toothed or toothless.

FLOWERHEADS: Sunflower-like, 2–3.5 cm wide, with **6–8 bright yellow rays** and a **dark central disc**, often nodding at maturity; late summer.

FRUITS: Achenes, **with 4 stiff, barbed bristles at tip**; barbs point downward.

WHERE FOUND: In swamps, wet muddy fields, beaver meadows and shores where lowered water table has left exposed mud; from Nova Scotia to British Columbia, south to California, South Dakota and North Carolina.

NOTES: Water marigold (*Megalodonta beckii*, p. 184) has similar flowers and upper stem leaves, but it is an aquatic species with submerged, thread-like leaves. It grows in quiet ponds and slow-moving streams from southeastern British Columbia to Nove Scotia, south to Oregon, Missouri, and New Jersey. • **Purple-stemmed beggar's ticks** (*B. connata*) has simple leaves, and its achenes have 2–4 bristles with upward-pointing barbs near their bases. It grows in swamps and on shores, from Nova Scotia to northwestern Ontario, south to Kansas, Tennessee and Virginia. • **Devil's beggar's ticks** (*B. frondosa*) has leaves that are divided into 5 leaflets, achenes with 2 stiff bristles and no ray florets. It grows in damp, open areas, from Newfoundland to British Columbia, south to California, Louisiana and Virginia. • **Small beggar's ticks** (*B. discoidea*) is similar to devil's beggar's ticks, but its leaves are thinner and more membranous, and they have only 3 leaflets. It also has much smaller flowerheads (5 mm high vs. 10–15 mm high) and its achenes have 2 bristles with barbs pointing towards (rather than away from) their tips. Small beggar's ticks grows in wet places, especially peaty shores and often on floating or partially decayed logs, from Ontario to Nova Scotia, south to Texas and Alabama. • The achenes are consumed by waterfowl and songbirds but they do not make up a large portion of their diet. • The species name *cernua* means 'drooping,' and it refers to the nodding mature flowers. Beggar's tick seeds are called 'hitchhikers' because their barbs hook onto animals for seed dispersal.

GRASSES, SEDGES AND RUSHES

This section describes narrow-leaved plants from 3 families: grasses (Poaceae), sedges (Cyperaceae) and true rushes (Juncaceae). Not all of the wetland species in these large families could be treated in this guide, but the most common and/or ecologically important species are included. As with all complex, taxonomically difficult plant groups, it is wise to collect voucher specimens and to have your determination verified by a botanist, whenever correct identification is important.

The 3 families can be separated using the following characteristics:

Grass Family (Poaceae)
- *stems hollow, jointed, round*
- *leaves in 2 vertical rows*
- *leaf sheaths usually open*
- *flowers with scales (palea and lemma)*
- *fruit a grain (caryopsis).*

Sedge Family (Cyperaceae)
- *stems solid, not jointed, usually 3-sided*
- *leaves in 3 vertical rows*
- *leaf sheaths closed*
- *flowers with 1 scale*
- *fruit an achene.*

Rush Family (Juncaceae)
- *stems solid at nodes, not jointed, round*
- *leaves usually basal*
- *leaf sheaths open*
- *flowers 3-parted, usually with 6 scales*
- *fruit a capsule with many seeds.*

**spikelet
(1-flowered)**

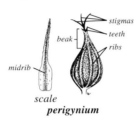

perigynium

flower

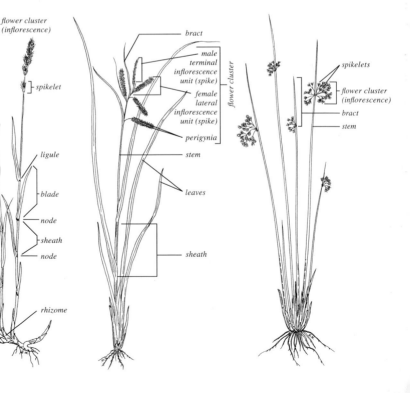

105

KEY TO GRASSES

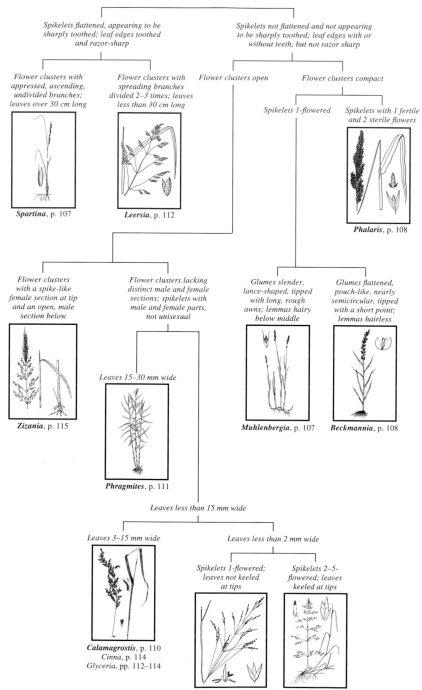

Spikelets flattened, appearing to be sharply toothed; leaf edges toothed and razor-sharp

Spikelets not flattened and not appearing to be sharply toothed; leaf edges with or without teeth, but not razor sharp

Flower clusters with appressed, ascending, undivided branches; leaves over 30 cm long

Flower clusters with spreading branches divided 2–3 times; leaves less than 30 cm long

Flower clusters open

Flower clusters compact

Spikelets 1-flowered

Spikelets with 1 fertile and 2 sterile flowers

Spartina, p. 107

Leersia, p. 112

Phalaris, p. 108

Flower clusters with a spike-like female section at tip and an open, male section below

Flower clusters lacking distinct male and female sections; spikelets with male and female parts, not unisexual

Glumes slender, lance-shaped, tipped with long, rough awns; lemmas hairy below middle

Glumes flattened, pouch-like, nearly semicircular, tipped with a short point; lemmas hairless

Leaves 15–30 mm wide

Zizania, p. 115

Muhlenbergia, p. 107

Beckmannia, p. 108

Phragmites, p. 111

Leaves less than 15 mm wide

Leaves 3–15 mm wide

Leaves less than 2 mm wide

Spikelets 1-flowered; leaves not keeled at tips

Spikelets 2–5-flowered; leaves keeled at tips

Calamagrostis, p. 110
Cinna, p. 114
Glyceria, pp. 112–114

Agrostis, p. 109

Poa, p. 109

MARSH TIMOTHY
MUHLENBERGIE AGGLOMÉRÉE • *Muhlenbergia glomerata*

GENERAL: Perennial grass; flowering stems 30–80 cm tall, solitary or in small clumps; from **scaly rhizomes**.

LEAVES: Hairless, 6–12 cm long, 2–6 mm wide; sheaths open to base.

FLOWER CLUSTERS: Narrow, almost cylindrical panicles, solitary, 2–11 cm long, 3–15 mm wide; spikelets 1-flowered, less than 3 mm long, congested, often purplish or bluish, turning pale yellow in late summer; **flower scales (lemmas) hairy at base, thin, flexible**, with a short stiff bristle (awn), **3-veined on back**; sometimes persisting through winter above snow.

FRUITS: Elongated, seed-like grains.

WHERE FOUND: Fen indicator, most common in water tracks of large peatlands and occasionally in floating mats on lakeshores; sometimes on dry sites; from Newfoundland to the Yukon Territory, south to Oregon, Minnesota and Connecticut.

NOTES: Marsh timothy could be confused with **timothy (*Phleum pratense*)**, but that grass has a denser, uninterrupted panicle, and it usually grows in upland habitats from Newfoundland to Alaska, south to California, Texas and Florida. • Some *Muhlenbergia* species are lightly browsed by deer and moose, though this has not been noted for marsh timothy. • The species name *glomerata* means 'aggregated into clusters.'

PRAIRIE CORD GRASS
SPARTINE PECTINÉE • *Spartina pectinata*

GENERAL: Perennial grass, with smooth, stout flowering **stems to 2 m tall**; from **long, tough, scaly rhizomes**.

LEAVES: Very **coarse**, 5–15 mm wide, to 60 cm long, flat or inrolled at edges; **sheaths hairless**, veined.

FLOWER CLUSTERS: Narrow panicles, 10–30 cm long, with **spikelets crowded along 1 side of each branch** in spike-like clusters; branches (spikes) alternate, erect, closely parallelling main stem, 4–8 cm long; **spikelets 1-flowered**; upper glume 8–12 mm long, tipped with a **4–10 mm awn**, rough with minute **barbs along keel**.

FRUITS: Seed-like grains, very small, 1 per spikelet.

WHERE FOUND: On wet lakeshores and riverbanks, often rooted in sand deposited among boulders by floodwater, also in marshes and wet prairies; from British Columbia and the southern Northwest Territories to Newfoundland, south to Oregon, Texas and North Carolina.

NOTES: The rhizomes of prairie cord grass form a shallow meshwork that is effective in impeding wave erosion and stabilizing sediments. • Prairie cord grass was formerly harvested in large quantities for hay in the Bay of Fundy area. This could have facilitated seed dispersal, and it may account for the scattered distribution of this species in Ontario. • This grass may be locally important as wildfowl food, but it is generally of low value for wildlife. • The genus name *Spartina* was derived from the Greek *spartine*, 'a cord,' because a species of this genus was used to make cords. The species name *pectinata* means 'comb-like,' and it refers to the arrangement of the spikelets on the stem.

SLOUGH GRASS
BECKMANNIE À ÉCAILLES UNIES • *Beckmannia syzigachne*

GENERAL: Pale green, perennial grass, 0.5–1 m tall, tufted, from fibrous roots.

LEAVES: Flat, 5–10 mm wide, rough, with **overlapping sheaths.**

FLOWER CLUSTERS: Elongate panicles with many, overlapping spikes up to 2 cm long; **spikelets flattened**, circular, with 1 flower concealed by **2 deep, keeled glumes 2–3 mm long**; mid- to late summer.

FRUITS: Oblong seed-like grains, falling free from spikelets.

WHERE FOUND: On edges of marshes and flooded river- and streambanks and in wet fields and ditches; from Nova Scotia to Alaska, south to California, Kansas, and Pennsylvania; possibly introduced in the east.

NOTES: Deer and moose browse slough grass, and it could be used as a 'beaver hay' grass by northern Ontario farmers. The seeds are a moderately good food for waterfowl. • The name *syzigachne* means 'scissor-like glumes'—a key identification feature.

REED CANARY GRASS
PHALARIS ROSEAU • *Phalaris arundinacea*

GENERAL: Robust, **leafy**, perennial grass, pale golden in autumn; flowering stems stiff, hollow, swollen at nodes, 70–150 cm tall, **in dense stands**; from **spreading rhizomes**.

LEAVES: Up to 20 mm wide, 10–20 cm long, **flat; ligules 2–6 mm long, membranous, blunt; sheaths open, hairless**, not overlapping.

FLOWER CLUSTERS: Dense panicles, about 10–20 cm long, with branches more or less spreading at maturity; **spikelets with 1 fertile and 2 sterile flowers**; outer flower scales (glumes) pointed, slightly keeled; fertile flower scales (lemmas) hairy near tip only, shiny below, with 2 tiny, silky-hairy scales (sterile lemmas) at base; mid-summer.

FRUITS: Seed-like grains, enclosed in shiny flower scales (lemmas) when shed.

WHERE FOUND: On muddy soil along lakes, rivers and streams and occasionally in marshes; most common in roadside ditches; usually introduced in Northern Ontario; from Newfoundland to Alaska, south to California, New Mexico and North Carolina.

NOTES: Canary grass (*P. canariensis*) comes from the Mediterranean, and it is grown commercially for domestic bird seed (canary seed). It is an annual species with a dense, oblong panicle, 2–4 cm long, and its outer flower scales (glumes) have strong, 1 mm wide keels. Canary grass is introduced occasionally at scattered locations, and it does not persist in northern Ontario. • Reed canary grass seeds are eaten by some songbirds, but this large grass generally has little value for wildlife, except as cover. • Some farmers grow reed canary grass as 'beaver hay' in moist fields. • The genus name *Phalaris* is from the Greek *phalaris*, 'a coot,' which in turn was derived from *phalos*, 'white,' in reference to its white head. It refers to the shiny fertile flower scales (lemmas).

TICKLE GRASS
AGROSTIDE SCABRE • *Agrostis scabra*

GENERAL: Tufted, perennial grass; flowering stems 30–90 cm tall, often slightly bent at nodes, fragile, hairless; from **fibrous roots**.

LEAVES: 1–2 mm wide, 10–30 cm long, **hairless**, rough on edges.

FLOWER CLUSTERS: Open panicles, 10–30 cm long, often **pinkish or purple**, with fine, wiry, rough branches; **spikelets single-flowered**, 2.5–3.2 mm long, **clustered at branch tips**; summer.

FRUITS: Seed-like grains, very small, 1 per spikelet.

WHERE FOUND: In roadside ditches and shrubby thickets and on damp rock outcrops, lakeshores and fen edges where soils dry out periodically; an early pioneer in wet areas that have been burned; from Newfoundland to Alaska, south to California, Texas and Florida.

NOTES: Two similar species also found in this area are **redtop** (*A. gigantea*), which spreads by rhizomes, and **creeping bent grass** (*A. stolonifera*), which spreads by stolons. Both can have red panicles, but they differ in that their spikelets are scattered from the bases to the tips of the panicle branches. These grasses are found in wet in damp sites across North America, south to California and Florida. • Tickle grass is browsed by deer and occasionally by moose. • The common name 'tickle grass' comes from the sensation you get when the panicle is lightly brushed against your bare leg or the bottom of your foot. Tickle grass is also called 'fire grass,' because of its reddish colour, and because it often colonizes burned areas.

FOWL MEADOW GRASS, FOWL BLUE GRASS
PÂTURIN PALUSTRE • *Poa palustris*

GENERAL: Perennial grass; flowering stems **purplish and curved at base, rooting at nodes, 50–150 cm tall**, loosely tufted; from stolons.

LEAVES: Narrow, 1–2 mm wide, **keeled like bow of boat at tips**; **ligules 3–5 mm long.**

FLOWER CLUSTERS: Loose, 8–30 cm long, open panicles; **branches in whorls of 4–5**; spikelets 2–4-flowered; flower scales (lemmas) **3-nerved, with a small tuft of cobwebby hairs at base, hairless and golden-bronze at tip**; mid-summer.

FRUITS: Seed-like grains, falling free at maturity.

WHERE FOUND: On shores, in wet fields, ditches, swamps and marshes and occasionally in fens; from Newfoundland to Alaska, south to California, Tennessee and North Carolina.

NOTES: There are many species of *Poa*, including the common **Kentucky blue grass** (*P. pratensis*) planted in lawns. Grasses of the genus *Poa* have leaves with boat-shaped (keeled) tips, and many have a tuft of cobwebby hairs at the base of their flower scales (lemmas). Fowl meadow grass is separated from the other species by its long ligule, large, open panicle and hairy, bronze-tipped flower scales (lemmas). • The leaves of fowl meadow grass are eaten by American coots in large quantities (providing up to 50% of their diet), and the seeds are eaten in small quantities by gamebirds and songbirds. Many mammals, including moose, deer, meadow voles and muskrats, feed on the leaves and seedheads of this grass.

CANADA BLUEJOINT
CALAMAGROSTIDE DU CANADA • *Calamagrostis canadensis*

GENERAL: Perennial grass, up to 1 m tall; stems densely clustered, with **swollen purple-blue joints**; from rhizomes.

LEAVES: 4–8 mm wide, flat, rough, slightly arching; **ligules 3–8 mm long, ragged near tips.**

FLOWER CLUSTERS: Open, **often purplish panicles**, 8–20 cm long; **spikelets single-flowered**, many, **with a tuft of straight hairs at base** almost as long as flower scales (lemmas); summer.

FRUITS: Plump, elliptic, seed-like grains.

WHERE FOUND: Most abundant in wet, open areas on mineral soil including beaver meadows, ditches and shores; also in swamps, fens and drier habitats; from Newfoundland to Alaska, south to California, Ohio, and Delaware.

NOTES: Canada bluejoint is extremely variable in size and colour, depending on its genetic make-up and environment. • **Northern reed grass** (*C. stricta* ssp. *inexpansa*, also known as *C. inexpansa*) is very similar to Canada bluejoint, but its flower clusters (panicles) are much narrower, with erect branches parallelling the main stalk. Also, the fine hairs at the base of the spikelet are only ²/₃ to ³/₄ as long as the flower scale (lemma), rather than equal to it, and it has rough rather than smooth lemmas. Northern reed grass grows on sand dunes and riverbanks and in rock crevices from Alaska to Newfoundland, south to California, New Mexico, Missouri, Michigan and Vermont. It can hybridize with Canada bluejoint grass. • Canada bluejoint grass could be confused with **reed canary grass** (***Phalaris arundinacea***, p. 108), which has wider leaves and very narrow, tight panicles, without tufts of long hairs in the spikelets. • Canada bluejoint is an important forage species harvested by farmers from wet fields in 'beaver hay.' Deer, moose and muskrats heavily graze the young shoots. The stems stand up well through the winter, and they provide cover for wildlife. • The genus name *Calamagrostis canadensis* means 'reed grass from Canada.'

C. canadensis   C. stricta

COMMON REED
ROSEAU COMMUN • *Phragmites australis*

GENERAL: Very large, perennial grass; flowering stems strong, smooth, 2–4 m tall, often forming large, dense colonies; from **strong, deep, creeping rhizomes and long, leafy stolons**.

LEAVES: Pennant-like, **flat, spreading, 1–3 cm wide**, 10–40 cm long; sheaths loose, overlapping.

FLOWER CLUSTERS: Many-branched panicle, 20–40 cm long; spikelets 3–7-flowered, 12–15 mm long, with many long, soft, white hairs from base giving a silky, feathery appearance, bisexual or male only; late summer.

FRUITS: Seed-like grains, **rarely produced**.

WHERE FOUND: In roadside ditches or marshes and in lakes (up to 2 m deep); occasionally in swamps and fens and sometimes spreading to colonize sand dunes; from Nova Scotia to the Northwest Territories, south to California, Texas, Indiana and Maryland.

NOTES: Common reed could be confused with **wild rice** (*Zizania palustris*, p. 115), but wild rice has dangling, hairless spikelets, and it is an annual grass with short roots that are easily pulled out. • Extensive, dense, colonies of common reed can replace other wetland plants. This large grass is of little value to wildfowl as food, but its rhizomes are a favourite food for muskrats. The stems and leaves provide summer and winter cover for wildlife. • Reeds are widely used by people throughout the world, for weaving mats, thatching roofs and providing other building materials. In Ojibway, *weenbushkoon* means 'reed,' and *minooshkoosawae* means 'he who cuts reeds for mats.' The genus name *Phragmites* is Greek, and it means 'growing in hedges,' referring to the hedge-like growth of this grass along ditches.

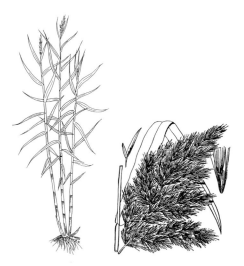

RICE CUT GRASS
LÉERSIE FAUX-RIZ • *Leersia oryzoides*

GENERAL: Light yellow-green, perennial grass; stems up to 150 cm long, branched, sprawling, rooting from nodes, forming dense mats; from slender, creeping rhizomes.

LEAVES: Rough-edged, flat, 15–30 cm long and 6–15 mm wide, abrasive on upper surface.

FLOWER CLUSTERS: Spreading, open panicles, up to 30 cm long; spikelets ascending, 1.5–2 mm wide; **flower scales (lemmas) keeled (V-shaped), fringed with hairs**; late summer.

FRUITS: Seed-like grains.

WHERE FOUND: On shores, especially along rich streambanks; from Nova Scotia to British Columbia, south to California, Texas and Florida.

NOTES: The dense mats of these rough leaves and stems can tear at clothing, and this can make walking difficult. In late summer, plants exposed to frequent flooding often produce flower clusters that remain hidden in the leaf sheaths, and these don't ripen until after the stem is dead.

NORTHERN MANNA GRASS
GLYCÉRIE BORÉALE • *Glyceria borealis*

GENERAL: Slender perennial grass; flowering stems 80–120 cm tall, often rooting from lower nodes; from rhizomes.

LEAVES: Leaves soft and spongy with large air spaces, flat, 2–5 mm wide, **sometimes floating; sheaths closed**.

FLOWER CLUSTERS: Slender panicle, 20–50 cm long; spikelets 7–13-flowered, **erect, cylindrical, slender, 1–2 cm long**, longer than their stalks; mid-summer.

FRUITS: Seed-like grains, 1–1.2 mm long.

WHERE FOUND: Often rooted in shallow water up to 60 cm deep, also on wet, muddy shores; from Newfoundland to Alaska, south to California, South Dakota and New England.

NOTES: Northern manna grass is sometimes confused with **wild rice (*Zizania palustris*,** p. 115), which grows in similar habitats and has a similar form, but wild rice produces much larger grains, and it has separate male and female spikelets. • The seeds of northern manna grass have been used as soup thickener and made into flour. **Caution:** see note about ergot fungus under wild rice (p. 115). • The leaves of this grass have a 'non-wettable' surface (i.e. water beads and rolls off).

TALL MANNA GRASS
GLYCÉRIE GÉANTE • *Glyceria grandis*

GENERAL: Perennial grass, up to 1.5 m tall; flowering stems usually in clusters, stout, spongy at base; from rhizomes.

LEAVES: Flat, 8–12 mm wide, 18–30 cm long; sheaths closed.

FLOWER CLUSTERS: Spreading panicle, **up to 40 cm long**, with many branches nodding at tips; spikelets 4–6.5 mm long, 5–9-flowered; flower scales (lemmas) purplish, with 7 parallel, raised ribs, **appearing corregated**.

FRUITS: Small, seed-like grains; summer.

WHERE FOUND: In shallow water, on shores and in ditches; most common on clay soils; from Newfoundland to Alaska, south to Arizona, Iowa and Virginia.

NOTES: This is our largest manna grass. • **Long manna grass** (*G. melicaria*) grows in moist rich woods from Ontario (where it is rare) to Nova Scotia, south to Tennessee and North Carolina. Its spikelets are smaller than those of tall manna grass (4 mm vs. 5–6 mm long), and they have fewer flowers (3–4 vs. 4–7). Its panicles are very narrow and erect, with branches that parallel the main stem, quite unlike the spreading, nodding panicles of tall manna grass. • Dense stands of manna grass along streams stabilize the soil, and they also provide excellent food and cover for wildlife.

RATTLESNAKE MANNA GRASS
GLYCÉRIE DU CANADA • *Glyceria canadensis*

GENERAL: Erect, perennial grass; flowering stems usually solitary, 60–100 cm tall; from rhizomes.

LEAVES: Flat, 3–8 mm wide, 15–35 cm long; sheaths closed.

FLOWER CLUSTERS: Loose, spreading panicles, 10–30 cm long, **nodding at branch tips**; **spikelets** 5–12-flowered, swollen, 5–8 mm long, **4–8 mm wide**, at tips of branches; **flower scales (lemmas) with visible (but not raised) nerves**, often purplish.

FRUITS: Small, seed-like grains; mid-summer.

WHERE FOUND: In shallow water, on shores and occasionally in fens; often abundant in beaver meadows; from Newfoundland to Saskatchewan, south to Illinois, Tennessee and South Carolina.

NOTES: The common name refers to the spikelets which resemble tiny rattlesnake rattles. • The Greek name *Glyceria* means 'sweet,' in reference to the seeds, which are said to be sweet and edible when fresh.

FOWL MANNA GRASS
GLYCÉRIE STRIÉE • *Glyceria striata*

GENERAL: Perennial grass; flowering stems slender, 50–120 cm tall, tufted; from rhizomes.

LEAVES: Blades flat, long, 2–5 mm wide; sheaths closed.

FLOWER CLUSTERS: Open panicle, 10–20 cm long, **with branches ascending, but nodding at tips**; spikelets green or purplish, 2–2.5 mm wide, 2.5–4.5 mm long, shorter than their stalks; flower scales (lemmas) with 7 parallel, **raised ribs, appearing corregated**; mid-summer.

FRUITS: Small, seed-like grains.

WHERE FOUND: On shores and in hardwood swamps, rich conifer swamps, ditches and fens; from Newfoundland to Alaska, south to California, Texas, and Florida.

NOTES: This species is highly variable, depending on growing conditions. • The seeds of fowl manna grass are a moderately good food source for waterfowl.

DROOPING WOODREED
CINNA À LARGES FEUILLES • *Cinna latifolia*

GENERAL: Slender, hairless, perennial grass, 60–120 cm tall; **loosely tufted** from rhizomes.

LEAVES: Limp, flat, spreading at right angles to stem, 15–25 cm long, 10–15 mm wide, rough to touch; ligules membranous, hairy, 5–10 mm long.

FLOWER CLUSTERS: Open, nodding panicles, 15–30 cm long, many-flowered, with clusters of slender, spreading branches; **spikelets 1-flowered**, 3–4 mm long.

FRUITS: Narrow, seed-like grains, very small, 1 per spikelet.

WHERE FOUND: On organic and mineral soils in richer swamps, damp woods, upland conifer and aspen mixedwoods and clearings and on roadsides and moist rock faces; grows well in shade, but may also become very abundant in sunlit disturbed areas, such as cutovers; from Alaska to Newfoundland, south to California, New Mexico and North Carolina.

NOTES: Drooping woodreed could be confused with **Canada bluejoint grass (*Calamagrostis canadensis*, p. 110)**, but that grass is often a deeper purple, and its flower scales (lemmas) have a tuft of white hairs at their base. • The green leaves can be burned on low fires to produce mosquito- (and camper-!) repelling smoke. • The genus name *Cinna* was derived from *kinni,* the Greek name for an unknown grass. The species name *latifolia* is from the Latin latus, 'broad,' and *folium,* 'leaf,' in reference to the wide leaves.

WILD RICE
ZIZANIE DES MARAIS • *Zizania palustris*

GENERAL: Robust, **annual, aquatic grass,** up to 2 m tall; from short roots; **easily uprooted.**

LEAVES: Flat, 10–60 cm long, up to 1.5–4 cm wide, soft, floating on water in spring, but lifting as plant emerges.

FLOWER CLUSTERS: Large panicle with many wide-spreading branches near base and erect (broom-like) branches near tip; spikelets 1-flowered, male or female; **male spikelets** 1.5–2 mm thick, **hanging (like little lanterns) from lower branches; female spikelets on club-shaped stalks on upper branches,** their flower scales (lemmas) firm and tough, straw-coloured and tinged red to purple, with 1–7 cm long bristles (awns) and with stiff bristles in furrows only; late summer.

FRUITS: Cylindrical, seed-like grains, **up to 1.5 cm long, slender,** dark brown or black, enveloped by long-awned flower scales (lemmas).

WHERE FOUND: In shallow lakes and quiet streams with loose, organic bottoms; from Nova Scotia to Manitoba, south to Texas and Florida.

NOTES: In the spring, young floating leaves resemble those of a **bur-reed** (*Sparganium* spp., pp. 162–64). • **Water oats** (*Z. aquatica),* also called **southern wild rice,** is very similar to wild rice, and sometimes these 2 species are considered varieties of the same species. They grow in similar habitats, but water oats is generally larger (about 2–3 m vs. 1 m tall), with wider (over 1 cm wide) leaves and wider, less densely clustered spikes. Lemmas on female florets are thin and papery, whitish, dull and slightly roughened with minute, sparse, stiff hairs. The male florets are narrow (less than 1.5 mm wide). Water oats tends to remain green until it is killed by frost in the autumn, whereas wild rice leaves and roots begin to die as soon as the grain is ripe, before the end of the growing season. Water oats is less common, and it is generally more southern than wild rice. It grows in marshes from Saskatchewan to Nova Scotia, south to Texas and Florida. • Wild rice is an annual grass, and therefore it reproduces each year by seed. • A **poisonous fungus** called ergot (*Claviceps* sp.) grows on the grains of many grasses, including wild rice. Infected grains can be distinguished by their pink or purplish to blackish colour, and they should not be eaten. • Wild rice is a very important food for waterfowl and marsh birds. Songbirds, such as red-winged blackbirds and bobolinks, feed heavily on wild rice. • In Ojibway, wild rice is called *minomin,* 'the

good seed.' It is still used extensively by native people across North America as a food, usually cooked with native herbs and wild meats. Wild rice is traditionally harvested by bending the stalks over a canoe and knocking off the seeds with a stick.

Z. aquatica

KEY TO THE SEDGE FAMILY (CYPERACEAE)

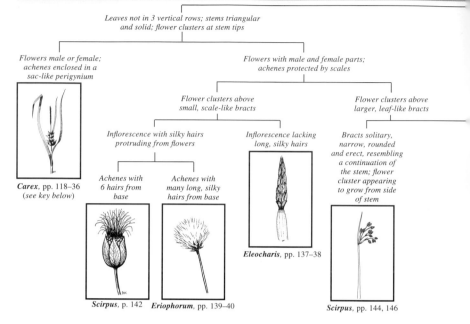

Leaves not in 3 vertical rows; stems triangular and solid; flower clusters at stem tips

Flowers male or female; achenes enclosed in a sac-like perigynium

Flowers with male and female parts; achenes protected by scales

Flower clusters above small, scale-like bracts

Flower clusters above larger, leaf-like bracts

Inflorescence with silky hairs protruding from flowers

Inflorescence lacking long, silky hairs

Bracts solitary, narrow, rounded and erect, resembling a continuation of the stem; flower cluster appearing to grow from side of stem

Carex, pp. 118–36
(*see key below*)

Achenes with 6 hairs from base

Achenes with many long, silky hairs from base

Eleocharis, pp. 137–38

Scirpus, p. 142   **Eriophorum**, pp. 139–40

Scirpus, pp. 144, 146

KEY TO THE SEDGES (*CAREX* SPP.)

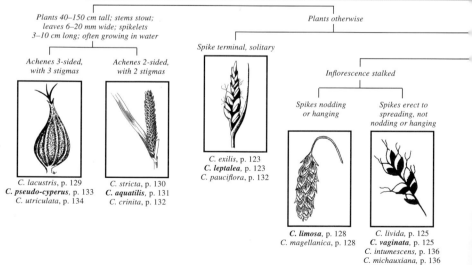

Plants 40–150 cm tall; stems stout; leaves 6–20 mm wide; spikelets 3–10 cm long; often growing in water

Plants otherwise

Spike terminal, solitary

Inflorescence stalked

Achenes 3-sided, with 3 stigmas

Achenes 2-sided, with 2 stigmas

Spikes nodding or hanging

Spikes erect to spreading, not nodding or hanging

C. lacustris, p. 129
C. pseudo-cyperus, p. 133
C. utriculata, p. 134

C. stricta, p. 130
C. aquatilis, p. 131
C. crinita, p. 132

C. exilis, p. 123
C. leptalea, p. 123
C. pauciflora, p. 132

C. limosa, p. 128
C. magellanica, p. 128

C. livida, p. 125
C. vaginata, p. 125
C. intumescens, p. 136
C. michauxiana, p. 136

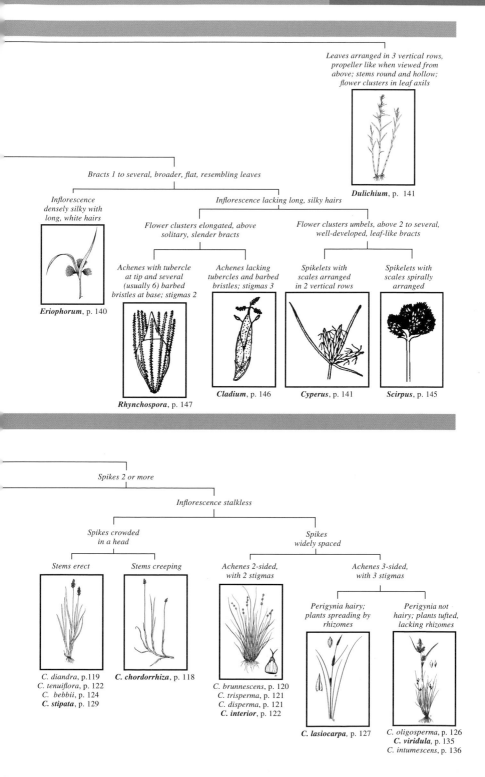

Leaves arranged in 3 vertical rows, propeller like when viewed from above; stems round and hollow; flower clusters in leaf axils

Dulichium, p. 141

Bracts 1 to several, broader, flat, resembling leaves

Inflorescence densely silky with long, white hairs

Inflorescence lacking long, silky hairs

Flower clusters elongated, above solitary, slender bracts

Flower clusters umbels, above 2 to several, well-developed, leaf-like bracts

Eriophorum, p. 140

Achenes with tubercle at tip and several (usually 6) barbed bristles at base; stigmas 2

Achenes lacking tubercles and barbed bristles; stigmas 3

Spikelets with scales arranged in 2 vertical rows

Spikelets with scales spirally arranged

Rhynchospora, p. 147

Cladium, p. 146

Cyperus, p. 141

Scirpus, p. 145

Spikes 2 or more

Inflorescence stalkless

Spikes crowded in a head

Spikes widely spaced

Stems erect

Stems creeping

Achenes 2-sided, with 2 stigmas

Achenes 3-sided, with 3 stigmas

Perigynia hairy; plants spreading by rhizomes

Perigynia not hairy; plants tufted, lacking rhizomes

C. diandra, p.119
C. tenuiflora, p. 122
C. bebbii, p. 124
C. stipata, p. 129

C. chordorrhiza, p. 118

C. brunnescens, p. 120
C. trisperma, p. 121
C. disperma, p. 121
C. interior, p. 122

C. lasiocarpa, p. 127

C. oligosperma, p. 126
C. viridula, p. 135
C. intumescens, p. 136

PARTS OF A SEDGE

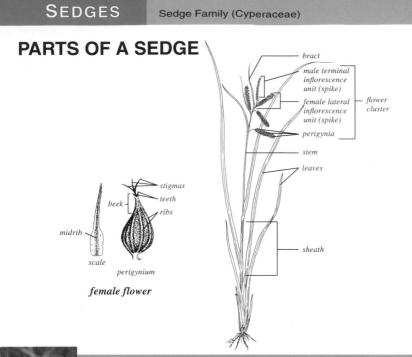

bract
male terminal inflorescence unit (spike)
female lateral inflorescence unit (spike)
perigynia
stem
leaves
flower cluster

stigmas
teeth
ribs
beek
midrib
scale
perigynium

female flower

sheath

CREEPING SEDGE
CAREX À LONG RHIZOME • *Carex chordorrhiza*

GENERAL: Small, perennial sedge; **old stems appear segmented, spread horizontally** sending up vertical shoots from nodes; vertical stems 10–30 cm tall, few produce fruit.

LEAVES: Sparse, 1–5 cm long, 1–2 mm wide.

FLOWER CLUSTERS: Crowded terminal clusters of 3–8 spikes, with male flowers at tip and female flowers at base of each.

FEMALE FLOWERS: 1–5 per spike; perigynia strongly veined, plump, 2.5–3.5 mm long and almost as wide; achenes lens-shaped, with 2 stigmas; scales brownish, equalling the perigynia; mature in early summer, before most other sedges are in flower.

WHERE FOUND: In fens; from Newfoundland to Alaska, south to British Columbia, Iowa and Vermont

NOTES: Creeping sedge is one of the best fen indicators in northern Ontario. It is never found in bogs. • Creeping sedge is easily overlooked because it grows low to the ground, and it blends in with other sedges such as **poor sedge** (*C. magellanica*, p. 128) and **candle lantern sedge** (*C. limosa*, p. 128). It is easily confused with candle lantern sedge if the flower cluster is absent. Both species have horizontal, segmented stems, but creeping sedge has noticeably 3-ranked leaves, is yellow-green at the base rather than deep brown, and it lacks felt-covered roots. • The genus name *Carex* means 'cutter,' and it refers to the sharp, saw-toothed leaf edges of most sedges. The species name *chordorrhiza* means 'slender rhizome,' and it refers to the thin stems that grow horizontally under the moss.

LESSER PANICLED SEDGE
CAREX DIANDRE • *Carex diandra*

GENERAL: Slender, perennial sedge, 30–100 cm tall; flowering stems sharply angled; in dense tussocks from fibrous roots.

LEAVES: 1–3 mm wide, shorter than flowering stems; **sheaths reddish-brown-dotted.**

FLOWER CLUSTERS: Dense heads of many, dark brown spikes; spikes small, stalkless, with male flowers at tip and female flowers at base.

FEMALE FLOWERS: Perigynia 2.4–3.0 mm long, egg-shaped, **dark brown, flattened at tip into a finely-toothed beak;** achenes lens-shaped, with 2 stigmas; scales brownish, equalling perigynia; mid-summer.

WHERE FOUND: In wet meadows, in fens and on lakeshores; from Newfoundland to Alaska, south to California, Nebraska and New Jersey.

NOTES: The distinctive dark brown flower clusters rise above those of **wire sedge** (*C. lasiocarpa*, p. 127), which is often found with lesser panicled sedge. • Lesser panicled sedge could be confused with several other sedges, but all of them can be easily separated in the field after a close examination. • **Bear sedge** (*C. arcta*) has male flowers at the base of its spikes, and it is always pale green. It is found in swamps and moist forests from New Brunswick to southern Alaska, south to California, Michigan and New England. • **Awl-fruited sedge** (*C. stipata*, p. 129) has leaves that are 4–10 mm wide. • **Prairie sedge** (*C. prairea*) has copper-tinged leaf sheaths and usually branched flower clusters. It is found from Nova Scotia to southern British Columbia, south to Nebraska, Ohio and New Jersey. **Fox sedge** (*C. vulpinoidea*) has straw-coloured (rather than brown or black) perigynia, the scales of its female flowers are awned (rather than awnless) and the inner side of its leaf sheaths have distinct cross-wrinkles. Fox sedge is fairly common in wet, open areas including lakeshores, riverbanks, ditches, meadows and woodland edges, from southern British Columbia to Newfoundland, south to Arizona, Texas and Florida. • The species name *diandra* means 'stamens in pairs.'

C. vulpinoidea

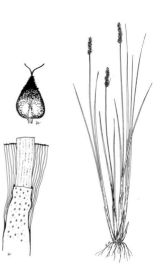

BROWNISH SEDGE
CAREX BRUNÂTRE • *Carex brunnescens*

GENERAL: Slender, perennial sedge; flowering stems 15–70 cm tall, stiff, erect, much taller than leaves; densely tufted from short rootstalk.

LEAVES: Soft, flat (sometimes channelled), 1–3 mm wide.

FLOWER CLUSTERS: Narrow, 2–6 cm long cluster of 4–8 well-spaced spikes; spikes 4–6 mm long, egg-shaped to rounded, **bisexual**, 4–10 flowered, with **female flowers at tip**; June–August.

FEMALE FLOWERS: Perigynia 2–2.7 mm long, 1–1.5 mm wide, faintly nerved, **covered with minute dots** (puncticulate), loose-spreading at maturity; **scales yellow or brownish**, shorter than perigynia; achenes lens-shaped, with 2 stigmas.

WHERE FOUND: In fens, rich swamps and wet woods; from Alaska to Newfoundland, south to Oregon, Utah, Minnesota and North Carolina.

NOTES: Silvery sedge (*C. canescens*) is very similar to brownish sedge, but its spikes have more (10 to many) flowers, and its leaves are bluish green with a whitish bloom (glaucous). It grows in fens, swamps, conifer forests and thickets and along stream and pond edges from Alaska to Newfoundland, south to California, Indiana and Virginia. • The species name *brunnescens* was derived from the French *brun*, 'brown,' in reference to the brown spikes of this species.

C. canescens

C. canescens

SOFT-LEAVED SEDGE
CAREX DISPERME • *Carex disperma*

GENERAL: Perennial sedge; **flowering stems 10–40 cm tall**, solitary or loosely tufted from rhizomes.

LEAVES: Flat, soft, narrow, 1–2 mm wide, shorter than flowering stems.

FLOWER CLUSTERS: 2–5 **well-separated, few-flowered spikes**; spikes with 1–4 female flowers at base and 1–3 male flowers at tip; lowermost bract less than 2 cm long.

FEMALE FLOWERS: Perigynia 2–3 mm long, **wide-spreading**; achenes thick, almost round, with 2 stigmas; scales thin, papery, ²/₃ as long as perigynia; early summer.

WHERE FOUND: In black spruce or cedar swamps and treed fens; able to tolerate dense shade under conifers in moist conditions; from Newfoundland to Alaska, south to California, Minnesota and Pennsylvania.

NOTES: Three-fruited sedge (*C. trisperma*, below) is similar to soft-leaved sedge, but its lowermost bract is 2–4 cm long, and its female flowers are borne at the tip of each spike. • The species name *disperma* means '2-seeded,' and it refers to the spikes, which usually have 2 flowers.

THREE-FRUITED SEDGE
CAREX TRISPERME • *Carex trisperma*

GENERAL: Small, perennial sedge; **stems very slender and weak, reclining or spreading**, 30–75 cm long, loosely tufted; from thin rhizomes or stolons.

LEAVES: Narrow, 1–2 mm wide, lax; many dead leaves usually at base of plant.

FLOWER CLUSTERS: Groups of 1–3 well-separated spikes; spikes 1–5-flowered, with male flowers at base and female flowers at tip; **slender, 2–4 cm long bract** below lowest spike, several times longer than spike.

FEMALE FLOWERS: 1–5 per spike; perigynia 2.4–4 mm long, with many fine nerves and a smooth beak; achenes lens-shaped, with 2 stigmas; scales translucent with a green midrib, shorter than perigynia.

WHERE FOUND: In treed bogs and conifer swamps; from Newfoundland to Northwest Territories, south to Minnesota, Illinois, and Virginia.

NOTES: Three-fruited sedge often grows intermixed with, and could be confused with **soft-leaved sedge** (*C. disperma*, above), which has shorter (less than 2 cm long) bracts. • Several other sedges also have well-separated spikes: **brownish sedge** (*C. brunnescens*, p. 120) has more (3–9) flowers per spike, and it is commonly found in swamps; **silvery sedge** (*C. canescens*) has more than 10 flowers per spike and more than 5 spikes, and it is commonly found in open, wet habitats from Newfoundland to Alaska, south to California, Ohio, and Virginia; **bristle-stalked sedge** (*C. leptalea*, p. 123) has a similar habitat and general appearance, but it has only 1 spike per stem. • The species name *trisperma* means 'three-fruited,' and it refers to the perigynia, which are usually in groups of 3.

SPARSE-FLOWERED SEDGE
CAREX TÉNUIFLORE • *Carex tenuiflora*

GENERAL: Wiry, perennial sedge; flowering stems sparse, 20–60 cm tall, gently arching at maturity; from yellow-brown rhizomes.

LEAVES: Shorter than flowering stems, 1–2 mm wide.

FLOWER CLUSTERS: Congested heads of 2–4 stalkless spikes, with **female flowers at tip** and **male flowers at base** of each.

FEMALE FLOWERS: Less than 10 per spike; perigynia whitish to pale brown or pale green, oval, 2.8–3.5 mm long, nerved; achenes almost filling perigynia, lens-shaped, with 2 stigmas; scales same size as perigynia, white with a green midrib; mid-summer.

WHERE FOUND: In fens and swamps and on peaty lakeshores; from Newfoundland to Alaska, south to British Columbia, Minnesota and Maine.

NOTES: Sparse-flowered sedge can be easily overlooked, because it is often loosely intermixed with other sedges, such as **three-fruited sedge** (*C. trisperma*, p. 121) and **soft-leaved sedge** (*C. disperma*, p. 121). • The species name *tenuiflora* means 'slender flower,' referring to the long, slender, few-flowered stems.

INLAND SEDGE
CAREX CONTINENTAL • *Carex interior*

GENERAL: Slender, perennial sedge; flowering stems wiry, 20–90 cm tall, **tufted**; from fibrous roots.

LEAVES: 1–3 mm wide, from basal third of stem, shorter than stem.

FLOWER CLUSTERS: Congested or well-separated groups of **3–5 small, few-flowered, stalkless spikes** with **male flowers at base** and female at tip, or with male flowers only.

FEMALE FLOWERS: Few, crowded, **wide-spreading, giving a starry appearance**; perigynia pear-shaped, with several veins on back and few veins or veinless on lower side, 2–3 mm long, extending past brown scales; **beak saw-toothed, no more than ¹/₃ the length of perigynia;** achenes lens-shaped, with **2 stigmas**; early summer.

WHERE FOUND: In fens, streams, wet meadows and cedar or spruce and tamarack swamps and on lakeshores; from the Yukon to Newfoundland, south to northern Mexico, Kansas and Delaware.

NOTES: Star sedge (*C. echinata*) and sterile sedge (*C. sterilis*) are similar to inland sedge, but their perigynia have longer beaks (more than ¹/₃ the length of the perigynia), with stiffer, longer teeth. The perigynia of star sedge are lance-shaped, whereas those of sterile sedge are pear-shaped. Star sedge can be found in acidic habitats, from Newfoundland to northern Ontario, south to California, Illinois and North Carolina. Sterile sedge can be found in fens and on shores, from Newfoundland to Saskatchewan, south to Tennessee and Pennsylvania.

STARVED SEDGE, COAST SEDGE
CAREX MAIGRE • *Carex exilis*

GENERAL: Perennial sedge; flowering stems 20–70 cm tall, taller than leaves, **forming tussocks**; from fibrous roots.

LEAVES: Slender, inrolled, usually shorter than flowering stem.

FLOWER CLUSTERS: Terminal spikes single (rarely 2 or more), 8–25 cm long, with several **empty scales** at base, usually with female flowers at tip and male at base, but sometimes male at tip or unisexual.

FEMALE FLOWERS: Up to 25 per spike, **spreading**; perigynia 2.5–5 mm long, saw-toothed on beak; scales reddish brown, with thin transparent edges; mid-summer.

WHERE FOUND: In rich open fens, particularly in water tracks of patterned peatlands; from Newfoundland to Lake Superior, south to Michigan, New York, and Delaware.

NOTES: Starved sedge often grows in association with **tufted clubrush** (*Scirpus cespitosus*, p. 142).

BRISTLE-STALKED SEDGE
CAREX À TIGES GRÊLES • *Carex leptalea*

GENERAL: Wiry, perennial sedge; flowering stems densely clustered, 15–30 cm tall; from **dense mats of rhizomes**.

LEAVES: 0.7–1.2 mm wide, wiry, shorter than flowering stems.

FLOWER CLUSTERS: Terminal spikes solitary, with male flowers at tip and a **single, hairlike bract** at base equal to or shorter than spike.

FEMALE FLOWERS: Ascending, pressed together, overlapping, 1–10 at base of spike; perigynia 2.4–4.5 mm long; achenes 3-sided, with 3 stigmas; mid-summer.

WHERE FOUND: In conifer swamps, treed fens, black ash swamps and open, wet areas in forests, such as trails and clearings; from Newfoundland to Alaska, south to California, Texas and Florida.

NOTES: This sedge is easily recognized by its delicate nature and its solitary spikes. • The name *leptalea* is Greek for 'thin' or 'delicate,' in reference to the thin, delicate stems and leaves.

BEBB'S SEDGE
CAREX DE BEBB • *Carex bebbii*

GENERAL: **Tufted**, perennial sedge; flowering **stems erect, slender, sharp-angled** and rough to touch near tip, 20–75 cm tall, taller than leaves; in **dense clumps** from short rhizomes.

LEAVES: Soft, flat, 2–5 mm wide, light green, **often yellowish; lowest leaves reduced to scales,** and well-developed leaves well above plant base.

FLOWER CLUSTERS: Dense, oblong clusters, 1–3 cm long, **greenish to straw-coloured** (brownish at maturity), with 5–10 stalkless spikes; spikes broadly ovoid to rounded, dense with many flowers, 4–10 mm long, **bisexual, with female flowers at tip**; June–August.

FEMALE FLOWERS: Perigynia **narrowly egg-shaped**, 2.4–3.9 mm long, 1.1–1.5 mm wide (up to 3 times as long as wide), **narrowly winged** along edges, tapering to a flattened, **2-toothed beak; scales pale to brownish**, shorter than perigynia; achenes lens-shaped, with 2 stigmas.

WHERE FOUND: On wet, sandy shores and streambanks and in meadows, ditches, cedar and tamarack swamps and clearings in damp woods; from southeastern Alaska to Newfoundland, south to Oregon, Colorado and New Jersey.

NOTES: Bebb's sedge is one of over 2 dozen species of the Ovales group of sedges, which are notoriously difficult to identify, even when in flower. These sedges generally have crowded, stalkless, bisexual spikes with female flowers at their tips, lens-shaped achenes and flattened, winged perigynia. In this group, the broom sedges are generally larger than Bebb's sedge, with plants reaching 1 m in height and perigynia 4–6.5 mm in length. • **Pointed broom sedge (*C. scoparia*)** has perigynia 4–5.5 mm long (2.3–3.5 times as long as wide), its main leaves are 1–3 mm wide, and its glossy brown or straw-coloured spikes have pointed tips. It is found on wet sandy shores of lakes, ponds and streams and in marshy grasslands, open swamps and wet meadows from southern British Columbia to Newfoundland, south to Oregon, New Mexico, Arkansas and Florida. • **Blunt broom sedge (C. *tribuloides*)** has perigynia 4–5 mm long and 1–1.3 mm wide (3.5–5 times as long as wide), its main leaves are 3–7 mm wide, and its dull, green to brown spikes have blunt, rounded tips. It is found in clearings, wet meadows, rich swamps and moist deciduous forests from Nebraska and Michigan to Nova Scotia, south to Oklahoma, Louisiana and Florida. • Bebb's sedge was named in honour of Michael Schuck Bebb, a botanist of the 1800s who studied willows. The species name *scoparia* comes from *scoparius*, which means 'broom-like.'

C. scoparia

SHEATHED SEDGE
CAREX ENGAÎNE • *Carex vaginata*

GENERAL: Perennial sedge; flowering stems erect, 20–60 cm tall, forming extensive, loose clumps; from long rhizomes.

LEAVES: Usually shorter than flowering stems, yellow-green, M-shaped in cross-section, 2–5 mm wide.

FLOWER CLUSTERS: Elongate groups of 2–4 widely separated, more or less spreading spikes, with 1 strongly stalked **male terminal spike and 2–4 female spikes below**; **spikes short-stalked above, long-stalked below**; lowest **bract with a loose, 1–2 cm long sheath**.

FEMALE FLOWERS: Loosely overlapping, 8–30 per spike; perigynia 3.5–5 mm long; achenes 3-sided, with 3 stigmas; scales purplish brown, shorter and narrower than perigynia; mid- to late summer.

WHERE FOUND: Usually in rich conifer (cedar or black spruce) swamps, also along streambanks; from Newfoundland to Alaska, south to Minnesota, Michigan and Vermont.

NOTES: This sedge could be confused with **livid sedge** (*C. livida*, below), but that species grows in fens, and its bracts have shorter sheaths. • The species name *vaginata* means 'sheathing,' and it refers to the long loose sheaths of the bracts.

LIVID SEDGE
CAREX LIVIDE • *Carex livida*

GENERAL: Stiffly erect, perennial sedge, with a **greyish-blue, powdery coating**; flowering stems slender, 20–50 cm tall, in small tufts; from **long rhizomes**.

LEAVES: Of varying lengths, lowermost shortest, uppermost often longer than flowering stems, **V- or M-shaped** in cross-section, 1–3 mm wide.

FLOWER CLUSTERS: Group of 2–4 erect spikes, with a 1–2 cm long male terminal spike (rarely with a few female flowers at base) and 1–2 short-stalked, 1–2 cm long, female spikes below.

FEMALE FLOWERS: Perigynia 3–5 mm long, with a bluish-white, waxy coating (glaucous); achenes 3-sided, with 3 stigmas, plump, filling the perigynia; scales shorter than perigynia, purple, with green stripe down centre.

WHERE FOUND: A fen indicator; from Newfoundland to Alaska, south to California, Minnesota and New Jersey.

NOTES: Similar sedges that grow in the same habitat as livid sedge are **candle lantern sedge** (*C. limosa*, p. 128) and **poor sedge** (*C. magellanica*, p. 128), but those species have nodding spikes on long stalks. • **Buxbaum's sedge** (*C. buxbaumii*) also has leaves with a bluish-white bloom, short-stalked to stalkless female spikes, light green perigynia and dark scales with green midribs. However, in Buxbaum's sedge the terminal spike always has female (rather than male) flowers at its tip. Buxbaum's sedge grows in meadows and fens and on shores from Alaska to Newfoundland, south to California, Colorado, Arkansas and North Carolina.

C. buxbaumii

125

FEW-SEEDED SEDGE
CAREX OLIGOSPERME • *Carex oligosperma*

GENERAL: Slender, sparsely fruited, perennial sedge; **flowering stems stiff**, brownish at base, 40–100 cm tall; from rhizomes and yellow to rust brown roots.

LEAVES: Shorter than flowering stems, 1–3 mm wide, **stiff, smooth, edges rolled in toward midrib, rounded in cross-section.**

FLOWER CLUSTERS: Elongated clusters of 2–4 spikes, usually with **a long-stalked, male spike at tip** and **1–3 widely separated**, stalkless **female spikes below**; lowermost bract long, slender, leaf-like.

FEMALE FLOWERS: 3–15 per spike; perigynia somewhat inflated, 4–7 mm long, **smooth, shiny, hairless**, strongly veined, abruptly beaked; achenes 3-sided, with 3 stigmas.

WHERE FOUND: Our most common sedge in bogs and poor fens; occasional along nutrient-poor, peaty lakeshores; from Newfoundland to northern Alberta, south to Minnesota, Ohio and Pennsylvania.

NOTES: Few-seeded sedge could be confused with **wire sedge** (*C. lasiocarpa*, p. 127), but wire sedge has reddish (rather than brown) stem bases, and its leaves are angular (rather than rounded) in cross-section.

• Few-seeded sedge is one of the few species that prefer acidic, nutrient-poor bog environments to richer fen conditions. It is most abundant in open bog drainage tracks. It colonizes mainly by its vigorous rhizomes, which play an important part in sedge mat formation. • The species name *oligosperma* means 'few-seeded.'

WIRE SEDGE
CAREX À FRUITS TOMENTEUX • *Carex lasiocarpa*

GENERAL: Perennial, slender sedge, 30–100 cm tall; from scaly rhizomes and white to drab-brown roots.

LEAVES: Longer than flowering stems, arching, narrow, 1–2 mm wide, **wire-like,** folded along midrib, angular; sheaths filamentous and **red-tinged at base.**

FLOWER CLUSTERS: Elongated groups of 2–5 erect, well-separated spikes, with **1–2 male terminal spikes** (2–6 cm long) and **1–3 stalkless female spikes** (1–4 cm long) **at base.**

FEMALE FLOWERS: Perigynia **densely woolly** (use a hand lens), 3–4 mm long, tipped with 2 stiff teeth; achenes with 3 stigmas; scales narrow, with brownish-purple edges; mid-summer.

WHERE FOUND: In fens and peaty marshes; the most common sedge of floating mats in northern Ontario; from Newfoundland to Alaska, south to Washington, Ohio and New Jersey.

NOTES: Wire sedge is a distinctive sedge, easily separated from **candle lantern sedge** (*C. limosa*, p. 128)*,* **livid sedge** (*C. livida*, p. 125) and **poor sedge** (*C. magellanica*, p. 128), all of which grow in similar habitats. Its extensive rhizome system forms the framework of most floating fen mats. • **Woolly sedge** (*C. lanuginosa*) is very similar to wire sedge, but it has flattened, rather than inrolled leaves. It grows on riverbanks and lakeshores and in marshes, fens and wet meadows from British Columbia to Newfoundland, south to California, Texas and Virginia. • Wire sedge could be confused with **few-seeded sedge** (*C. oligosperma*, p. 126), which also has long, wiry leaves, but few-seeded sedge has hairless perigynia, brown (rather than red) stem bases and leaves that are rounded (rather than angular) in cross-section. • The species name *lasiocarpa* means 'hairy fruit.'

CANDLE LANTERN SEDGE, MUD SEDGE
CAREX DES BOURBIERS • *Carex limosa*

GENERAL: Perennial sedge; flowering stems hairless, slightly reddish near base, turning dark brown in late summer, 20–60 cm tall; from **slender, sprawling, segmented rhizomes; roots covered with yellow-brown felt.**

LEAVES: Few, bluish green, shorter than stem, 1–3 mm wide.

FLOWER CLUSTERS: Groups of 2–4, well-separated spikes, with **1 long-stalked, 15–30 cm long, male terminal spike** and **1–5 female spikes hanging** on slender stalks below.

FEMALE FLOWERS: Perigynia dull powdery-green, strongly flattened; achenes 3-sided, with 3 stigmas; **scales** brownish, **shorter than or equal to perigynia and just as broad**; mid-summer.

WHERE FOUND: A fen indicator, in hollows and shallow pools; from Newfoundland to Alaska, south to California, Saskatchewan and Delaware.

NOTES: Candle lantern sedge is one of the easiest sedges to recognize at a distance because of its dangling spikes, which hang like candle lanterns. It often grows in association with **poor sedge** (*C. magellanica*, below), **creeping sedge** (*C. chordorrhiza*, p. 118), **livid sedge** (*C. livida*, p. 125) and **wire sedge** (*C. lasiocarpa*, p. 127) • The only other sedge that grows in fens and has spikes like this is poor sedge, which has weaker rhizomes, longer, narrower, purple-black female scales and shorter male spikes. • Its strongly rhizomatous nature makes candle lantern sedge an important component of floating sedge mats.

POOR SEDGE
CAREX CHÉTIF • *Carex magellanica*

GENERAL: Perennial sedge; flowering stems 20–70 cm tall, reddish brown with many old leaves near base, **loosely clustered**; from short rhizomes; roots **covered with yellowish-brown felt.**

LEAVES: Flat, green, 1–3 mm wide, shorter than stem.

FLOWER CLUSTERS: Elongated clusters of 2–4 spikes with a **5–12 mm long, male terminal spike** and 1–3 **stalked, nodding** or sometimes slightly erect, **female spikes below**.

FEMALE FLOWERS: Perigynia flattened, dull powdery-green; achenes 3-sided, with 3 stigmas; **scales** dark purple-black, **narrower and longer than perigynia**; mid-summer.

WHERE FOUND: Poor fens and coniferous (black spruce, cedar or tamarack) swamps, and in rocky crevices along Lake Superior shores; from Newfoundland to Alaska, south to Washington, Minnesota and New England.

NOTES: This species was formerly known as *C. paupercula*. • Poor sedge often grows intermixed with **candle lantern sedge** (*C. limosa*, above), **creeping sedge** (*C. chordorrhiza*, p. 118), **livid sedge** (*C. livida*, p. 125) and **wire sedge** (*C. lasiocarpa*, p. 127). • Candle lantern sedge could be confused with poor sedge, but it has more extensive rhizomes, longer male spikes and shorter broader, brown female scales.

AWL-FRUITED SEDGE
CAREX STIPITÉ • *Carex stipata*

GENERAL: Robust, perennial sedge; flowering stems densely clustered, **stout, sharply angled, triangular**, 30–100 cm tall.

LEAVES: Rough, flat, 4–8 mm wide, usually cross-puckered; sheaths with wavy ridges.

FLOWER CLUSTERS: Crowded, 2–10 cm long heads of many, small, stalkless, few-flowered spikes, with male flowers at tip and female flowers at base of each spike.

FEMALE FLOWERS: Perigynia short-stalked, 4–5 mm long, wide-spreading, nerved, with 2 distinct teeth at tip; achenes lens-shaped, with 2 stigmas; scales shorter than the perigynia; mid-summer.

WHERE FOUND: In ditches, marshes, rich swamps and other wetlands; from Newfoundland to southern Alaska, south to California, Texas, and Florida.

NOTES: Most sparrows eat sedge fruits mainly in the autumn, winter and spring, but the swamp sparrow feeds on perigynia throughout the year, and these fruits make up almost 25% of its diet. The tree sparrow's diet consists of 5–10% sedge seeds during the autumn and spring, when it is migrating through northern Ontario. • The species name *stipata* means 'stalked,' and it refers to the perigynia, which have a short stalk.

LAKEBANK SEDGE
CAREX LACUSTRE • *Carex lacustris*

GENERAL: Robust, perennial sedge; flowering stems 50–150 cm tall, sturdy, **purplish or reddish at base**; from stout, scaly rhizomes.

LEAVES: M-shaped in cross-section, rough, hairless, 8–15 mm wide, usually longer than stem; lower sheaths becoming a network of fibres.

FLOWER CLUSTERS: Elongated clusters of 4–8 well-separated spikes, with **2–4 male terminal spikes**, and 2–4 cylindrical, 3–10 cm long, female spikes at base.

FEMALE FLOWERS: Perigynia beaked, 4–7 mm long, firm, somewhat leathery, 12–25-nerved; achenes 3-sided, with 3 stigmas; scales purple-tinged with **short, stiff bristle (awn)**; mid-summer.

WHERE FOUND: Often in shallow water up to 50 cm deep, in marshes and ditches, on shores and at edges of floating fen mats; from Newfoundland to Saskatchewan, south to Idaho, South Dakota and Virginia.

NOTES: Lakebank sedge could be confused with **beaked sedge (*C. utriculata*, p. 134)** but that sedge has membranous perigynia with fewer (7–10) nerves, and its leaves are narrower (2–10 mm), and they are usually V-folded, rather than M-folded. • Lakebank sedge seeds are eaten by waterfowl and songbirds. The stems and leaves remain through winter, and they provide spawning habitat for pike and muskellunge the following spring.

TUSSOCK SEDGE
CAREX RAIDE • *Carex stricta*

GENERAL: Erect, perennial sedge; flowering stems sharply 3-angled, with **many bladeless sheaths at base**, 40–140 cm tall, **often forming large tussocks**; from rhizomes.

LEAVES: 3–6 mm wide, rarely longer than flowering stems; **lowest leaves reduced to bladeless sheaths**.

FLOWER CLUSTERS: Clusters of 2–7 spikes with 1 (sometimes 2) stalked, **male terminal spikes** and 2–5, erect, cylindrical, stalkless, female spikes below.

FEMALE FLOWERS: Many, crowded; perigynia green to reddish brown, elliptic, 1.5–3.5 mm long, face upwards; achenes lens-shaped, with 2 stigmas; scales blunt, as long as perigynia, reddish brown or purple, with a pale centre; mid-summer.

WHERE FOUND: In fens, marshes, swamps, ditches and wet meadows and on shores; from Nova Scotia to Manitoba, south to Colorado, Texas and Florida.

NOTES: Tussock sedge can be confused with **water sedge** (*C. aquatilis*, p. 131), which grows in similar habitats, but water sedge does not have bladeless leaf sheaths, and its stem bases are surrounded by dried up leaves from the previous year. • **Hayden's sedge** (*C. haydenii*) is very similar to tussock sedge, and both were once considered varieties of the same species. They are difficult to separate without their basal parts. Hayden's sedge has short, ascending rhizomes, rather than long, horizontal rhizomes, and its lower leaf sheaths are smooth and not split into filaments, whereas those of tussock sedge are rough and red-brown, and they split to form a feather-like network. Hayden's sedge also has wider (broadly egg-shaped to rounded) perigynia that are much shorter than their sharply pointed, spreading scales. Hayden's sedge grows in marshes, wet meadows and wet open woods from Ontario (where it is rare) to New Brunswick, south to Nebraska, Missouri and New Jersey. • Rails and snipes nest in the tussocks formed by tussock sedge. • Tussock sedge is an important component of 'beaver hay' harvested from wet fields. This species was formerly harvested for insulation and packing material in ice houses, and for rug making. • The species name *stricta* means 'tall and straight.'

WATER SEDGE
CAREX AQUATIQUE • *Carex aquatilis*

GENERAL: Robust, perennial sedge, pale green; flowering stems 30–150 cm tall, sharply 3-angled above; in dense tufts from **long, scaly rhizomes.**

LEAVES: Bluish white, hairless, 4–7 mm wide; **lower sheaths often red-tinged.**

FLOWER CLUSTERS: Elongate group of several, well-separated spikes; spikes 1.5–7 cm long, **erect, corn-cob-like**, often stalkless; **uppermost 2–3 spikes male; lower 3–5 spikes mainly female** (may have male flowers at tips).

FEMALE FLOWERS: Many, densely packed; perigynia 2–3 mm long, flattened, short-beaked; scales dark-edged; achenes lens-shaped, with 2 stigmas; summer.

WHERE FOUND: Along streambanks and lakeshores and in marshes, wet fields, fens and ditches; from Newfoundland to Alaska, south to California, Ohio, and New Jersey.

NOTES: Water sedge can be confused with **beaked sedge** (*C. utriculata*, p. 134), with which it often grows intermixed, but beaked sedge can be distinguished by its 3 stigmas, its broader, yellow-green leaves and its larger, strongly beaked perigynia. • Water sedge is also very similar to **tussock sedge** (*C. stricta*, p. 130), but that species has bladeless leaf sheaths at the base of its stems. • **Lenticular sedge** (*C. lenticularis*) is very similar to water sedge, though less common. It has narrower (1–3 mm wide) leaves and single male spikes (rather than 2 or more), and it lacks the long stolons found in water sedge. It can also be recognized by close examination of its perigynia, which have a few sharp, elevated nerves on each face, and which are covered with minute granules not found in water sedge. Lenticular sedge grows on wet, sandy and rocky shores of lakes, rivers and beaver ponds from British Columbia to Newfoundland, south to California, Colorado, Michigan and Massachusetts. • Sedge seeds are an important food for both waterfowl and songbirds.

Water sedge makes up the bulk of 'beaver hay' cut from wet meadows. • The species name *aquatilis* means 'growing in water.'

C. lenticularis

131

FEW-FLOWERED SEDGE
CAREX PAUCIFLORE • *Carex pauciflora*

GENERAL: Slender, perennial sedge; **flowering stems few**, usually less than 20 cm tall; from **long slender rhizomes**.

LEAVES: 1–2 mm wide, shorter than flowering stems.

FLOWER CLUSTERS: Solitary terminal spikes, up to 1 cm long, bractless, with **male flowers at tip** and female flowers at base.

FEMALE FLOWERS: 1–6 per spike; perigynia narrow, 6–8 mm long, widely reflexed at maturity; achenes 3-sided, with 3 stigmas; scales pale brown, shorter than perigynia, soon falling off.

WHERE FOUND: In open or treed bogs under black spruce and tamarack; also on peatmoss hummocks in fens; from Newfoundland to Alaska, south to Washington, Minnesota and Connecticut.

NOTES: Few-flowered sedge often grows in association with **few-seeded sedge** (*C. oligosperma*, p. 126) • A similar species, **bristle sedge** (*C. microglochin*), has shorter stems and smaller (3–5 mm long) perigynia. In Ontario, it is restricted to the Hudson Bay and James Bay lowlands, but its range extends from northern Newfoundland to Alaska, south to central Manitoba and James Bay. • The species name *pauciflora* means 'few-flowered.'

FRINGED SEDGE
CAREX CRÉPU • *Carex crinita*

GENERAL: Robust, hairless, perennial sedge; flowering stems 50–150 cm tall, slightly taller than leaves; from stout rhizomes.

LEAVES: Flat, 4–10 mm wide, rough-edged; **upper leaves largest**.

FLOWER CLUSTERS: Nodding clusters of **long, stalked spikes**, with 1–3 male terminal spikes and 2–6 female spikes below; male spikes to 5 cm long, sometimes with a few female flowers towards base; **female spikes narrowly cylindrical**, 3–10 cm long, on **slender stalks**, often with a few male flowers at the tip.

FEMALE FLOWERS: Perigynia green, 2–3 mm long, **thin, inflated**, rounded, **nearly beakless; scales tipped with a long, rough-toothed point**, brown with green midrib, narrower than perigynia but 2–3 times longer; achenes lens-shaped, with 2 stigmas.

WHERE FOUND: In rich swamps, damp thickets and woods and along shorelines; from Ontario (and possibly southern Manitoba) to Newfoundland, south to Louisiana and Georgia.

NOTES: Nodding sedge (*C. gynandra*) is very similar to fringed sedge, and these 2 sedges were once classified as varieties of the same species. They are most easily separated by examining their leaf sheaths, which are rough with stiff hairs on nodding sedge and smooth on fringed sedge. Nodding sedge grows in swamps, wet woods, marshes and ditches and along the shores of rivers and ponds from Ontario to Newfoundland, south to Louisiana and Georgia.

CYPERUS-LIKE SEDGE
CAREX FAUX-SOUCHET • *Carex pseudo-cyperus*

GENERAL: Robust, perennial sedge; **flowering stems stout,** sharply 3-angled, rough, 30–100 cm tall; in clumps from short, stout rhizomes.

LEAVES: Rough, large, longer than stems, 5–15 mm wide.

FLOWER CLUSTERS: Elongate clusters of **3–6 nodding, long-stalked spikes,** with **1 shorted-stalked, male terminal spike** and 2–5 cylindrical, 3–7 cm long, female spikes below; **bracts longer than flowering stems.**

FEMALE FLOWERS: Many, crowded; **perigynia pointing backward, 4–6 mm long, leathery,** many-ribbed, with long teeth at tip; achene 3-sided, with 3 stigmas; **scales tipped with rough, stiff bristles** (awns); late summer.

WHERE FOUND: In swamps (especially cedar swamps), marshes and wet ditches and on shores; Newfoundland to Lake Superior and Alberta, south to North Dakota, Indiana and Pennsylvania.

NOTES: Cyperus-like sedge often grows with a similar species, **porcupine sedge** (*C. hystericina*), which is quite common on lakeshores and streambanks and in wet ditches from Newfoundland to southern British Columbia, south to California, Texas and Virginia. However, porcupine sedge has strong, stolon-like rhizomes, and its perigynia are thinner-walled, inflated (nearly round in cross-section) and ascending to spreading at maturity. • **Bristly sedge** (*C. comosa*) is very similar to cyperus-like sedge, but it is often larger, with stems up to 1.5 m tall and leaves to 16 mm wide. Its perigynia have much longer beaks (equalling the body), and they are tipped with long (1.2–2.3 mm), spreading teeth rather than shorter (0.5–1 mm long), more or less straight teeth. Bristly sedge grows in the shallow water of marshes, cedar swamps and fens and on the shores of lakes, pools and streams from Minnesota to Nova Scotia, south to Louisiana and Florida, and from southern British Columbia, south to California and Idaho. • **Sallow sedge** (*C. lurida*) is another large wetland sedge with large, ribbed perigynia and long (1.5–7 cm) female spikes. However, its spikes are short-stalked to stalkless (rather than slender-stalked) and its perigynia are membranous and inflated (rather than leathery and somewhat flattened). Also, sallow sedge has narrower (4–6 mm

wide) leaves and the slender beaks of its perigynia end in minute teeth (less than 0.7 mm long). It grows in marshes, swamps, wet woods and ditches and along the shores of lakes and streams from Minnesota to Nova Scotia, south to eastern Mexico and Florida. • Marsh birds eat many sedge seeds year round as part of their diet. The sora's diet can be 10–25% sedge seeds. • The species name *pseudo-cyperus* means 'false cyperus,' and it refers to the gross similarity of this plant and plants of the genus *Cyperus*. There are many differences between these 2 genera, the most obvious being that *Carex* species have perigynia, and *Cyperus* species do not.

BEAKED SEDGE, BOTTLE SEDGE
CAREX ROSTRÉ • *Carex utriculata*

GENERAL: Robust, perennial sedge; **flowering stems 50–120 cm tall**, blunt-edged; from rhizomes that grow deep into mud.

LEAVES: Longer than flowering stems, folded lengthwise, 4–10 mm wide, with a bluish-white, waxy coating; **sheaths with prominent cross-markings between veins; ligules as long as wide**.

FLOWER CLUSTERS: Elongated clusters of 4–8 spikes with female spikes at base and male terminal spikes; female spikes 2–10 cm long, up to 1.5 cm thick, **elongated, cylindrical, well-separated**, short-stalked or stalkless, more or less erect; male spikes 2–7 cm long.

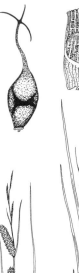

FEMALE FLOWERS: Many, dense, in 8–12 rows; perigynia **inflated**, wide-spreading, 4–7 mm long, up to 4 mm wide, **strongly nerved**, beaked, tipped with short teeth; achene 3-sided, with 3 stigmas; scales greenish, shorter than perigynia; mid-summer.

WHERE FOUND: In marshes, swamps, ponds, fens and ditches and on wet, muddy streambanks; sometimes in up to 20–30 cm of standing water; from Newfoundland to Alaska, south to California, Ohio and Delaware.

NOTES: This species is sometimes called *C. rostrata*. • Beaked sedge can be confused with certain other bladder sedges. • **Tuckerman's sedge** (*C. tuckermanii*) is recognized by its large, 5–7 mm wide perigynia and its deeply notched achenes. It grows in old beaver meadows and other wet sites from New Brunswick to Lake Superior, south to Iowa, Ohio, and New Jersey. • **Retrorse sedge** (*C. retrorsa*) is distinguished by its 7–12 mm long perigynia, its long lowermost bract (3 times as long as the flower cluster) and its short-stalked, male spikes, which are not elevated above the female spikes. It grows in alder thickets and hardwood swamps from Newfoundland to the Northwest Territories, south to Oregon, South Dakota and New Jersey. • **Inflated sedge** (*C. vesicaria*) is very similar to beaked sedge, but it has fewer perigynia (6–8 rows), its ligules are twice as long as wide, and its stems are sharp-edged. Inflated sedge grows in fens and swamps. • Sedge seeds are often eaten by waterfowl, songbirds and ruffed grouse chicks.

C. tuckermanii

GREEN SEDGE
CAREX VERDÂTRE • *Carex viridula*

GENERAL: Bright green, hairless, perennial sedge; flowering stems slender, smooth, erect, 5–60 cm tall; tufted, lacking rhizomes; sometimes forming dense patches.

LEAVES: Usually **yellowish green**, usually channelled, 1–3 mm wide, **erect**, often longer than flowering stalks; sheaths hairless, loose.

FLOWER CLUSTERS: Crowded, 1–5 cm long clusters of **stalkless or short-stalked spikes**, with 1 male terminal spike and 2–10 female spikes below; male spike 5–15 mm long; female spikes oblong, 5–10 mm long; **lowest bract extending past uppermost spike.**

FEMALE FLOWERS: Horizontally spreading perigynia green to brownish green, strongly nerved, ovoid, narrowed at base, beaked at tip, 2.2–3.3 mm long, straight or nearly so; scales yellowish brown, shorter than perigynia; achenes 3-sided and nearly filling perigynia, with 3 stigmas.

WHERE FOUND: In calcium-rich areas, on sandy to rocky, often marly shores and in beach pools and interdunal swales and occasionally in fens; from southern Alaska to Newfoundland, south to California, New Mexico, North Dakota and New Jersey.

NOTES: This species is also known as *C. oederi* **var.** *viridula*.
• **Hidden-scaled sedge** (*C. cryptolepis,* sometimes included in *C. flava*) could be confused with greenish sedge but its leaves are 1.5–3.5 mm wide, and its perigynia are longer (3.2–4.5 mm) and usually strongly sickle-shaped (with at least the beaks conspicuously bent

downwards). The perigynia hide their scales, and they have smooth beaks. Hidden-scaled sedge grows on calcium-rich soil on sandy to marly or mucky shores of lakes or streams from Ontario to Newfoundland, south to Minnesota, Indiana and New Jersey. • **Yellow sedge** (*C. flava*) is very similar to green sedge, and it also grows in wet areas, but its leaves are wider (2.5–5 mm) and flat (rather than channelled). It also has larger (5–6.2 mm long) perigynia, many or most of which are strongly sickle-shaped (with at least their beaks bent downward). The shiny, brown to reddish scales of the mature female flowers are conspicuous, and the beaks of the perigynia are often very minutely and sparsely toothed near their tips. Yellow sedge grows in fens, marshes, cedar and tamarack swamps (often in marly conditions), wet meadows and ditches and on wet shores from Ontario to Newfoundland, south to Minnesota, Indiana and New Jersey. • The species name *viridula* comes from the Latin *viridis*, meaning 'green,' and it refers to the distinctive yellowish-green colour of this sedge. The species name *cryptolepis* means 'hidden scale,' and *flava* means 'yellow.'

135

MICHAUX'S SEDGE
CAREX DE MICHAUX • *Carex michauxiana*

GENERAL: Slender, yellowish, perennial sedge; stems erect, 20–60 cm tall; **densely tufted**, from short rhizomes.

LEAVES: Firm, 2–4 mm wide; sheaths hairless, loose.

FLOWER CLUSTERS: 2–10 cm long cluster of 3–5 short-stalked to stalk-less spikes, with **1 male spike at tip and 2–4 female spikes below**; lower spikes more widely spaced; **male spike 5–15 mm long, often hidden by female spikes**; female spikes ovoid to nearly round, 15–25 mm long, several- to many-flowered; **bracts extending past uppermost spike.**

FEMALE FLOWERS: Perigynia slender, narrowly lance-shaped, **yellow**, distinctly veined, 8–12 mm long, erect or spreading; scales brown, half as long as perigynia, blunt to pointed; achenes 3-sided, with 3 stigmas.

WHERE FOUND: In fens and on wet shores; from Ontario to Newfoundland, south to Michigan and Massachussetts.

NOTES: Michaux's sedge has disjunct populations in northeastern North America and eastern Asia. • **Long sedge** (*C. folliculata*) resembles Michaux's sedge, but its lax, flat leaves are wider (4–17 mm), its scales are tipped with long, rough-edged awns, and its male spikes are usually stalked, their tips projecting well above the female spikes. It grows in wet woods and cedar swamps from Wisconsin to Ontario (where it is rare) to Newfoundland, south to Louisiana and Georgia. • Michaux's sedge is named after Andre Michaux (1746–1803), a French botanist and collector who spent 10 years exploring in North America, where he collected over 60,000 plants for French herbaria.

BLADDER SEDGE
CAREX GONFLÉ • *Carex intumescens*

GENERAL: Robust, perennial sedge; flowering stems hairless, erect, 30–80 cm tall; usually tufted.

LEAVES: Soft, dark green, 4–8 mm wide, usually shorter than stem.

FLOWER CLUSTERS: Groups of 2–4 spikes, with 1 narrow, long-stalked, **male terminal spike** and 1–3 rounded, 1–3 cm long female spikes at base; lowermost bracts long, leaf-like, extending beyond flower cluster.

FEMALE FLOWERS: 1–12 per spike; perigynia beaked, inflated, **up to 16 mm long**, widely spreading, with a satiny lustre; achenes 3-sided, with 3 stigmas; scales narrower and shorter than perigynia.

WHERE FOUND: In black ash and alder swamps, from Newfoundland to Lake Superior and Lake Winnipeg, south to Texas and Florida.

NOTES: Hop sedge (*C. lupulina*, inset photo) also has large (13–20 mm long), inflated, ribbed, strongly beaked perigynia in short-stalked to stalkless spikes, but its female spikes are larger and more elongated (3–6 cm long and 2–3 cm thick), with many erect to slightly spreading perigynia. Hop sedge usually has long rhizomes (absent in bladder sedge), and its perigynia have beaks that are nearly as long (over 4.4 mm) as their bodies. It grows in wet woods, swamps, thickets, meadows and marshes from Minnesota to Nova Scotia, south to Texas and Florida. • The species name *intumescens* means 'swollen,' and it refers to the swollen fruits.

NEEDLE SPIKERUSH
ÉLÉOCHARIDE ACICULAIRE • *Eleocharis acicularis*

GENERAL: Perennial sedge, less than **10 cm tall**; flowering stems very slender, deep green, often have **reddish basal sheath, densely clustered, often forming mats in shallow water**; sterile stems often present; from thin, creeping rhizomes with fibrous roots.

LEAVES: Reduced to **bladeless** sheaths at base of stems.

FLOWER CLUSTERS: Solitary, 2–6 mm long spikes of bisexual flowers in axils of egg-shaped scales; **styles 3-branched, with an enlarged base** (tubercle).

FRUITS: 3-sided, pearly white achenes, with a cone-shaped cap (tubercle) at tip.

WHERE FOUND: In marshes and on lakeshores and marshy riverbanks; often submerged, forming mats on lake bottoms; from Newfoundland to Alaska, south to California, Mexico and Florida.

NOTES: Although common, this tiny plant is often overlooked. • There are several small species of spikerush with the same general appearance. This is a difficult group of plants, and identifications should be confirmed by careful examination of the flowers and mature fruits. • **Intermediate spikerush (*E. intermedia*) has 3-branched styles**, greenish or yellowish achenes and 10–30 cm long stems. It grows in wet places from Quebec to Lake Superior, south to Iowa, Tennessee and Maryland. • **Ovate spikerush (*E. ovata*)** has 2-branched styles, tubercles less than ²/₃ as wide as their achenes and purplish-brown scales. It grows on shores from Newfoundland to Lake Superior, south to Indiana and Connecticut. • **Blunt spikerush (*E. obtusa*)** has 2-branched styles, tubercles at least ²/₃ as wide as their achenes and reddish-brown scales. It grows on disturbed, muddy sites from Nova Scotia to British Columbia, south to California, Texas and Florida. • **Bright-green spikerush (*E. olivacea*)** is a tiny,

E. acicularis

mat-forming spikerush, with tufted plants growing from running rhizomes. Its flattened, thread-like flowering stems are 2–10 cm tall, and they bear 3–7 mm long spikes with blunt-tipped, egg-shaped scales and lens-shaped achenes. It grows on moist, sandy to muddy shores and exposed mud flats and occasionally in fen mats, from Minnesota to Ontario (where it is rare) to Nova Scotia, south to Florida. • The species name *acicularis* means 'needle-like,' and the stems of this tiny spikerush often appear to form mats of vertical needles.

E. obtusa

E. olivacea

E. ovata

E. intermedia

MARSH SPIKERUSH
ÉLÉOCHARIDE DE SMALL • *Eleocharis smallii*

GENERAL: Perennial sedge, 10–100 cm tall; **flowering stems rounded, in loose clusters**; **from thick rhizomes.**

LEAVES: Reduced to bladeless sheaths at base of stem.

FLOWER CLUSTERS: Solitary spikes of bisexual flowers in axils of lance-shaped, pointed scales; **styles 2-branched, with an enlarged base** (tubercle); 1 sterile flower at base of spike; summer.

FRUITS: 2-sided, yellow-brown achenes, nearly 2 mm long, with a large, bulb-like cap (tubercle) at tip.

WHERE FOUND: In most marshes and on most lakeshores and riverbanks, also occasional in wet fens; from Newfoundland to Manitoba, south to California, Texas and Virginia.

NOTES: This species is included in *E. palustris* by some taxonomists. • You need to examine the achenes (usually with a 10x lens) to positively identify spikerushes, and special attention should be paid to the size and shape of the tubercle and the number of style branches (2 or 3). Several other spikerushes with loosely clustered flowering stems and thick rhizomes are found in this region. • **Red-stemmed spikerush** (*E. erythropoda*, included in *E. palustris* by some) has 2-branched styles and 2–3 sterile scales at the base of each spike. It grows on wet soil from Quebec to Alberta, south to New Mexico, Tennessee and Virginia. • **Compressed spikerush** (*E. compressa*) has 2-branched styles and flattened flowering stems, and its achenes are covered with tiny bumps. It grows in fens from southern Quebec to British Columbia, south to Colorado, Texas and Georgia. • **Elliptic spikerush** (*E. tenuis*) has 2-branched styles and strongly 4–8-angled flowering stems, and its achenes are covered with tiny bumps. It grows on wet soil from Newfoundland to Alberta, south to Montana, Ohio and South Carolina. • **Few-flowered spikerush** (*E. pauciflora*) has 2-branched styles and slightly triangular to rounded flowering stems. It grows in fens from Newfoundland to the Yukon Territory, south to California, South Dakota Ohio and New Jersey. • Spikerushes are an important food for wild birds (especially waterfowl), which eat the stems and spikes. The rhizomes are also eaten by wild birds and by muskrats. • The genus name *Eleocharis* was derived from the Greek *helos*, 'a marsh,' and *charis*, 'grace.' Despite the common name, these are sedges, not rushes.

E. compressa

DENSE COTTONGRASS
LINAIGRETTE DENSE • *Eriophorum vaginatum*

GENERAL: Densely tufted, perennial sedge; flowering stems 20–60 cm tall, stiff, 3-sided, forming broad tussocks.

LEAVES: Basal leaves **1 mm wide**, several; stem leaves reduced to 1–2 loose, **inflated, bladeless sheaths**.

FLOWER CLUSTERS: Solitary terminal, cotton-ball-like spikelets, **lacking leafy bracts**; flowers bisexual, in axils of blackish scales with pale edges; early summer.

FRUITS: Achenes, 2.5–3.5 mm long, half as wide, 3-sided, **surrounded by many, long, white, silky hairs from base**.

WHERE FOUND: Often covering large expanses of open bogs, also in open conifer swamps; one of the few plants to prefer more acidic, nutrient-poor bog environments to richer fen conditions; from Newfoundland to Alaska, south to southern Alberta, Minnesota and New Jersey.

NOTES: This species is also known as *E. spissum*. • Two other cottongrasses with solitary spikelets and no floral bracts often share the same habitat as dense cottongrass.

• **Short-anthered cottongrass** (*E. brachyantherum*) has no rhizomes, its achenes are 3 times as long as wide, and its scales lack whitish edges. Also, its uppermost leaf sheath is slender (not inflated), and it is borne above the middle of the stem. It grows in wet places from Newfoundland to Alaska, south to southern British Columbia and Lake Superior. • **Rusty cottongrass** (*E. chamissonis*) has stems that grow singly (not in tussocks) from creeping rhizomes. It grows in wet, peaty sites from northern Newfoundland to Alaska, south to Oregon, Colorado and Minnesota. • Despite the common name, this is a sedge, not a grass. The genus name *Eriophorum* means 'wool-bearing,' and it refers to the cottony spikelets.

E. chamissonis

GREEN COTTONGRASS
LINAIGRETTE VERTE • *Eriophorum viridi-carinatum*

GENERAL: Perennial sedge; flowering stems stiff, 20–90 cm tall, several together or solitary; from slender rhizomes.

LEAVES: Flat (except inrolled at very tip), all bearing blades, 2–6 mm wide, mostly at base of stem.

FLOWER CLUSTERS: Groups of **3–30 cottony spikelets** nodding on slender stalks, with 2–3 long, leaf-like bracts at base; flowers bisexual, in axils of **blackish-green scales with prominent, pale midribs**; mid-summer.

FRUITS: Achenes, 2.5–3.5 mm long, 3-sided, with many, long, whitish, silky bristles from base.

WHERE FOUND: In fens and open conifer swamps; from Newfoundland to northern Alberta, south to southern British Columbia, Iowa and Ohio.

E. viridi-carinatum

NOTES: Several other species of cottongrasses with multiple spikelets and floral bracts are also found in this region. • **Rough cottongrass** (*E. tenellum*) has whitish bristles and a single bract that is shorter than the flower cluster, and the blade of the uppermost stem leaf is up to 2 mm wide, 3-sided, and equal to or longer than its sheath. Rough cottongrass grows in rich fens from Newfoundland to northern Ontario, south to Illinois, Michigan and New Jersey. • **Slender cottongrass** (*E. gracile*) has whitish bristles and a single bract that is shorter than the flower cluster, and the blade of the uppermost stem leaf up to 2 mm wide, 3-sided, and much shorter than its sheath. Slender cottongrass also grows in rich fens, and it is found from Newfoundland to Alaska, south to California, Minnesota and Delaware. • **Narrow-leaved cottongrass** (*E. angustifolium*) has several bracts that are longer than its flower cluster. Its spikelets have whitish bristles, and the midribs of its scales do not reach the scale tips. Narrow-leaved cottongrass grows in fens from Newfoundland to Alaska, south to Oregon, Illinois and New England. • **Tawny cottongrass** (*E. virginicum*) also has several bracts that are longer than its flower cluster, but its spikelets sometimes have tawny or coppery bristles, and each has sterile scales with no midrib at its base. Tawny cottongrass grows in bogs and poor fens from Newfoundland to Lake Superior, south to Minnesota, Tennessee and Georgia. It flowers in August, later than the other cottongrasses.

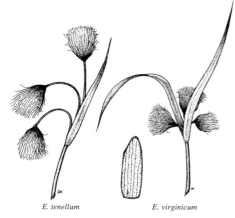

E. tenellum   *E. virginicum*

E. virginicum

THREE-WAY SEDGE
DULICHIUM ROSEAU • *Dulichium arundinaceum*

GENERAL: Erect, perennial sedge, **resembling a 3-armed airplane propeller from above**; flowering **stems hollow, jointed, thick, rounded** (often appear triangular because of 3-ranked leaves), 30–100 cm tall; from thick, spreading rhizomes.

LEAVES: Many, **in 3 vertical rows** (ranks), 2.5–7.5 cm long, 4–8 mm wide, reduced to bladeless sheaths on lower $^1/_3$ to $^1/_2$ of stem.

FLOWER CLUSTERS: Stalked, brownish clusters of **spikelets, in axils of upper leaves**; spikelets stalkless, linear, spreading, with 6–12 **flowers in 2 vertical rows; flowers bisexual**, with 3 stamens and a 2-branched style, in axils of lance-shaped scales; late summer.

FRUITS: Achenes, 2.5–3 mm long; 6–9 stiff, barbed **bristles** from base.

WHERE FOUND: In marshes, ponds and fens and on riverbanks; can form large beds in water less than 50 cm deep; from Newfoundland to British Columbia, south to California, Texas and Florida.

NOTES: Three-way sedge leaves have an interesting feature: when there are 2 adjacent plants on the same rhizome, one will have leaves spiralled clockwise, and the other will have leaves spiralled counter-clockwise. • Three-way sedge is not easily confused with other sedges, once you recognize its 3-ranked leaves. • Three-way sedge is sometimes eaten by waterfowl and muskrats, but it is not an important food source.

UMBRELLA-SEDGE, NUTGRASS
SOUCHET BIPARTITE • *Cyperus bipartitus*

GENERAL: Annual sedge; 10–40 cm tall; from **fibrous roots**.

LEAVES: Mainly **basal**; upper stem leaves forming a **whorl (involucre) at base of flower cluster**.

FLOWER CLUSTERS: Branching clusters (umbels) of linear, somewhat pointed spikes; spikes densely flowered, 8–20 mm long; flowers numerous, with male and female parts; **scales reddish brown** (especially **near base and** toward edge), overlapping and pressed together, thick, **almost leathery, slightly shiny**, blunt-tipped; style deciduous, often not conspicuous at maturity of spikelet; achenes oblong, dull, lens-shaped, with 2 stigmas.

WHERE FOUND: On wet, often marshy, sandy or muddy shores of ponds, lakes and streams and in wet meadows, ditches and swales; from California to Minnesota and Maine, south to Mexico and Georgia.

NOTES: This species is also known as *C. rivularis*. • Umbrella-sedge is difficult to distinguish from **two-stamened galingale** (*C. diandrus*), which sometimes grows with it in southern and central Ontario. Two-stamened galingale can be distinguished by the concentration of reddish-brown colour near the tips of its scales (rather than near the base) and by its more or less persistent style, which is conspicuous at maturity. • There are about 10 other species of *Cyperus* in Ontario, many of which are found only occasionally on sandy lakeshores in the southern part of the province. • Waterfowl occasionally eat the seeds of umbrella-sedge.

HUDSON BAY CLUBRUSH
SCIRPE HUDSONIEN • *Scirpus hudsonianus*

GENERAL: Perennial sedge; **flowering stems triangular, less than 40 cm tall,** loosely tufted; from tufted roots and slender rhizomes.

LEAVES: Stiff, 5–12 mm long, 5–12 mm wide, most reduced to bladeless sheaths near base of stem; upper 1–2 sheaths have a short blade.

FLOWER CLUSTERS: Solitary, 5–7 mm long spikelets, tipped with long, white, **silky hairs,** usually with a single, small bract at base; **flowers** bisexual, in axils of and **hidden by yellowish-brown scales.**

FRUITS: Brown achenes, 1–5 mm long, 3-sided, with 6 long, silky, white bristles from base.

WHERE FOUND: In fens and peaty ditches and on Lake Superior shores; from Newfoundland to Alaska, south to Montana, Minnesota and New England.

NOTES: Hudson Bay clubrush is distinguished from **dense cottongrass (*Eriophorum vaginatum*, p. 139)** by its smaller size and yellow-brown, rather than blackish, scales. • **Tufted clubrush (*S. cespitosus*, below)** is another small, tufted bulrush found in rich fens and on Lake Superior shores. It can be distinguished by its rounded flowering stems and smaller, non-silky spikelets. • Despite its common name, this is a sedge, not a rush.

TUFTED CLUBRUSH
SCIRPE GAZONNANT • *Scirpus cespitosus*

GENERAL: Perennial sedge; flowering stems slender, wiry, **rounded,** 10–40 cm tall, **forming dense tussocks.**

LEAVES: Most reduced to bladeless sheaths at base of stem; uppermost sheath with a slender, 4–6 mm long blade.

FLOWER CLUSTERS: Solitary spikelets, about 4 mm long, 2–4-flowered; flowers bisexual, in axils of pointed, brown scales; lowest scale (bract), slender-tipped, about as long as spikelet.

FRUITS: Achenes, 1.5–2 mm long, with **6 slightly longer, white, hair-like bristles** from base.

WHERE FOUND: In rich fens and on Lake Superior shores; from Newfoundland to Alaska, south to Oregon, Minnesota and North Carolina.

NOTES: Tufted clubrush is distinguished from **Hudson Bay clubrush (*S. hudsonianus*, above)** by is rounded, rather than triangular, stems and its shorter, less prominent, hair-like bristles. • Although it is primarily a rich fen species in northern Ontario, in eastern Canada, tufted clubrush grows in more acidic habitats, including bogs. At the southern edge of its range, it is restricted to areas influenced by the cold Lake Superior microclimate. • The species name *cespitosus* means 'tufted.'

TORREY THREE-SQUARE
SCIRPE DE TORREY • *Scirpus torreyi*

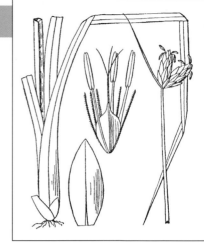

GENERAL: Perennial sedge; flowering **stems 40–100 cm tall, with 3 concave sides, solitary**; from slender, weak, creeping rhizomes.

LEAVES: Basal, light green, channelled, firm or lax, slender.

FLOWER CLUSTERS: Dense, 3–5 cm long **clusters (umbels) of 1–4 stalkless spikes;** spikes 10–15 mm long, ovoid to cylindrical, pointed; **bract solitary,** 3–15 cm long, blunt, erect, **appearing to be an extension of stem**; flowers with male and female parts and about 6 downwardly **barbed bristles**; **scales** egg- to lance-shaped, **chestnut brown, shiny, pointed**; achenes light brown, smooth, shiny, 3-sided, with 3 stigmas.

WHERE FOUND: In rich marshes and on shorelines; from southern Manitoba to New Brunswick, south to Missouri and Georgia.

NOTES: Two very similar species are also found in Ontario's wetlands. • **Common three-square** (*S. pungens*, also known as *S. americanus*) can be recognized by its scales, which have a slender bristle-tip projecting from a notched tip and a fringe of fine hairs. Its leaves are half as long as its stems. Common three-square grows on shores and in marshes, from Alaska to Newfoundland, south to Mexico and Florida.
• **Smith's three-square** (*S. smithii*) is a smaller (5–60 cm tall), tufted, annual sedge, with slender, rounded stems and leaves that are more than half as long as the stems. Its achenes are lens-shaped, with 2 stigmas, and the bristles at their bases have been reduced to 1 or 2 tiny remnants (or they are absent altogether). It grows on wet, sandy to mucky or peaty shores (especially where water levels have receded), from Minnesota to Quebec, south to Illinois and Virginia, but it is rare in Ontario. • Torrey three-square is named in honour of Dr. John Torrey (1796–1873), an eminent American botanist, co-author of the *Flora of North America* with Asa Gray. The common name, 'three-square,' refers to the 3-sided stems of these plants. Of the 3 species described here, common three-square is by far the most common in Ontario.

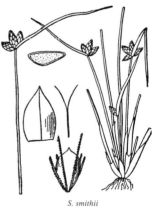

S. smithii

S. pungens

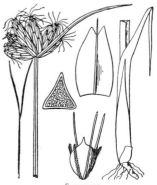

S. pungens

HARDSTEM BULRUSH
SCIRPE AIGU • *Scirpus acutus*

GENERAL: Emergent, perennial sedge; flowering stems rounded, 1–3 m tall, olive green, hard (not easily crushed between the fingers; from stout, scaly rhizomes.

LEAVES: Reduced to inconspicuous, bladeless sheaths at base of stem.

FLOWER CLUSTERS: Tight clusters (panicles) of spikelets that appear to grow from **side of stem**; spikelets egg-shaped, 5–10 mm long, 1–5 at tip of slender stalks; flowers bisexual, in axils of pale- or whitish-brown, orange-red-dotted scales.

FRUITS: Achenes, 1.5–2.5 mm long, brown to black, with 0–6 barbed bristles at base, in axils of (and hidden by) scales; mid-summer.

WHERE FOUND: One of the most common plants on sandy, wave-washed lakeshores and in sheltered bays and in ditches; from Newfoundland to Alaska, south to California, Texas and Georgia.

NOTES: Hardstem bulrush is very similar to **softstem bulrush** (*S. validus*), and these 2 species are sometimes considered subspecies of *S. lacustris*. Softstem bulrush can be distinguished by its light blue-green stems, which are soft (easily crushed between the fingers). Its spikelets are often in more open clusters (panicles) than those of hardstem bulrush, and they have rich orange-brown scales. Softstem bulrush grows in marshes and on shores from Newfoundland to British Columbia, south to California, Texas and Georgia • Two three-square bulrushes (with 3-sided stems) are found in this region. **Common three-square** (*S. pungens*, p. 143) has many spikelets, and its short, stiff, narrow leaves are less than half the length of the flowering stems. **Torrey three-square** (*S. torreyi*, p. 143) has many spikelets, and its leaves are narrow, stiff and more than half the length of the flowering stems, often overtopping them. • The Ojibway call bulrushes *anaukunushkoon*, and they use them for weaving mats and bedding.

S. acutus

S. validus

S. validus

WOOLGRASS
SCIRPE SOUCHET • *Scirpus cyperinus*

GENERAL: Perennial sedge; flowering stems triangular, up to 2 m tall, densely tufted; from tufted, fibrous roots and short rhizomes.

LEAVES: Ridged, very rough, 3–10 mm wide, over 30 cm long; sheaths closed.

FLOWER CLUSTERS : Loose, 15–30 cm long clusters (umbels) of many, rounded, nodding clusters of 6–12 spikelets; **spikelets brownish, 10 mm long, 2–4 mm wide, woolly at maturity**; flowers bisexual, in axils of and **hidden by scales**; several leaf-like bracts at base of flower cluster; late summer.

FRUITS: Whitish, 3-sided achenes, 0.7–1 mm long, with 6 long, **white to rust-coloured bristles** from base.

WHERE FOUND: In shallow marshes, beaver meadows and swamps; quickly colonizes soil in disturbed, wet cutovers and ditches; from Newfoundland to Manitoba, south to Oklahoma and North Carolina.

NOTES: Several other species of *Scirpus* in this region have spreading leaf-like bracts below the flower cluster. • **River bulrush** (*S. fluviatilis*) has spikelets that are 10–30 mm long and 6–12 mm wide. Its stems are strongly triangular and its rhizomes are thick and short. River bulrush grows in marshes (usually in shallow standing water) from New Brunswick to Alberta, south to California, Kansas and Virginia. • **Small-fruited bulrush** (*S. microcarpus*) has non-woolly spikelets that are about 10 mm long and 2–4 mm wide. Its flowering stems are loosely tufted, with reddish nodes. Small-fruited bulrush grows in marshes, ditches and wet meadows from Newfoundland to Alaska, south to California, Michigan and Virginia. • **Black bulrush** (*S. atrovirens*) also has non-woolly spikelets that are about 10 mm long and 2–4 mm wide. However, its flowering stems are densely tufted, and the scales of its spikelets are dark brown or black. Black bulrush grows in swamps and wet meadows from Newfoundland to Alberta, south to Oregon, Texas and Georgia.

S. fluviatilis   *S. microcarpus*

FLOATING BULRUSH, MERMAID'S HAIR
SCIRPE SUBTERMINAL • *Scirpus subterminalis*

GENERAL: Aquatic, perennial sedge; stems weak, 30–100 cm long, **floating or submerged**; from slender rhizomes.

LEAVES: Submerged, weak, grass-like, often forming large mats that gently wave in water currents.

FLOWER CLUSTERS: Solitary, stalkless spikelets, 6–13 mm long, slightly raised above water; flowers bisexual, in axils of (and hidden by) pointed, membranous scales, 1 leaf-like bract, resembling a continuation of stem; mid-summer.

FRUITS: 3-sided achenes, with slender beak at tip and 6 downward-barbed bristles from base.

WHERE FOUND: Usually in slow-moving streams, also in quiet lakes, rivers and ponds; from Newfoundland to Alaska, south to Oregon, Michigan and Georgia.

NOTES: Bulrush seeds provide valuable food for waterfowl and marsh birds. The diet of some coots may include up to 25% bulrush seeds. Songbirds occasionally eat these seeds, and the stems and roots are eaten by geese and muskrats. Blackbirds and marsh wrens build nests on bulrush stems. These plants also provide cover for muskrats, otters and raccoons, and their submerged leaves shelter fish and aquatic insects. • Despite their common name, bulrushes are sedges, not rushes. The genus name *Scirpus* is the classical Latin name for a rush, said to be derived from the Celtic word *sirs*, 'rushes.'

TWIGRUSH
MARISQUE À FEUILLES ENSCIE • *Cladium mariscoides*

GENERAL: Spiky, perennial sedge, turning dark brown in late summer; flowering stems smooth, rounded or multi-angled, stiff, slender, up to 100 cm tall; from spreading rhizomes.

LEAVES: 1–3 mm wide, almost round at tip.

FLOWER CLUSTERS: Branched, 5–30 cm long clusters (cymes) of 3–10 brown, scaly spikelets; 2 or more bracts at base of cluster; spikelets with 1 fertile flower (with 2 stamens and a 3-branched style) at tip and sterile flowers below; **flowers in axils of (and hidden by) flat, overlapping, spirally arranged, brown scales**; late summer.

FRUITS: Oval achenes, with fine, lengthwise stripes.

WHERE FOUND: In fens and on marshy lakeshores; from Newfoundland to Lake Superior, south to Minnesota, Alabama and Florida.

NOTES: The spreading rhizomes of twigrush are important in forming floating sedge mats. • The only plants that might be confused with twigrush are the bulrushes (*Scirpus* spp., p. 142–46), but those plants have many fertile flowers in each spikelet. • Little is known about the food value of twigrush to wildlife. The famous **sawgrass** (*C. jamaicense*) of the Florida Everglades, a closely related species, is an important duck food, and it is occasionally eaten by marsh birds and muskrats. • Despite the common name, this is a sedge, not a rush. The genus name *Cladium* means 'branched,' and it refers to the branched flower cluster.

WHITE BEAKRUSH
RHYNCHOSPORE BLANC • *Rhynchospora alba*

GENERAL: Pale green, perennial sedge; flowering stems very slender, usually less than 25 cm tall (but up to 70 cm tall); from short rhizomes.

LEAVES: 0.5–2.5 mm wide, stiff, often bristle-like, **shorter than flowering stem**.

FLOWER CLUSTERS: Flat-topped clusters of **4–5 mm long spikelets**, with several leaf-like bracts at base; flowers bisexual, in axils of (and hidden by) **white to pale brown scales**; style with an enlarged base (tubercle); mid-summer.

FRUITS: Achenes egg-shaped, with a large, persistent cap (tubercle) at tip, narrowed and with 8–14 long, downward-barbed bristles at base.

WHERE FOUND: Fen indicator; common in hollows and pools of open fens; from Newfoundland to Alaska, south to California, Minnesota and North Carolina.

NOTES: **Brown beakrush** (*R. fusca*) has brown scales and 6 or fewer perianth bristles with upward-pointing barbs. Brown beakrush grows in fens and on wet peat and sand from Newfoundland to Lake Superior, south to Michigan and Delaware.
• **Capillary beakrush** (*R. capillacea*) and **slender beakrush** (*R. capitellata*, also known as *R. glomerata*) have chestnut brown (rather than white) flower scales and 6 (rather than 9–15) bristles from the base of each achene. Slender beakrush has flat, slender (up to 3.5 mm wide) leaves and smooth, oblong achenes, whereas capillary beakrush has thread-like leaves (less than 0.5 mm wide), and its egg-shaped achenes are slightly roughened, with dark, more or less horizontal lines. Capillary beakrush grows in moist sandy or peaty shores, wet meadows, ditches and fens, from Alberta to Newfoundland, south to South Dakota, Tennessee and Virginia. Slender beakrush is found on wet sandy or stoney shores and in peaty beach pools, fens, marshes and meadow marshes from Wisconsin to Ontario (where it is rare) to Nova Scotia, south to eastern Texas and Florida.
• Beakrush achenes are eaten by waterfowl, but do not make up a large portion of their diet. • Despite its common name, white beakrush is a sedge, not a rush. The genus name *Rhynchospora* means 'beaked seed,' and it refers to the enlarged persistent cap (tubercle) on the seed. The species name *alba* means 'white,' and it refers to the whitish scales.

R. alba

R. capitellata

R. fusca

R. capillacea

147

CANADA RUSH
JONC DU CANADA • *Juncus canadensis*

GENERAL: Perennial rush; flowering stems stiff, up to 1 m tall, tufted; from thick, branching rhizomes.

LEAVES: Hairless, 1.5–2.5 mm wide, **rounded in cross-section, with cross-ridges** (can be felt by placing leaf on a flat surface and running a fingernail along the leaf surface).

FLOWER CLUSTERS: Spreading, terminal panicles of **dense**, 5–10-flowered, **spherical heads** up to 1 cm in diameter; **flowers** bisexual, with **6 green or brown, scale-like tepals**; late summer.

FRUITS: Capsules, 3.5–4.5 mm long, with **many, tiny seeds** that are tailed at both ends.

WHERE FOUND: In marshes with muddy or sandy shores and around small ponds in sand dunes; from Newfoundland to Lake Superior, south to Louisiana, Tennessee and Georgia.

NOTES: Two other rushes in our region have leaves with cross-ridges. **Short-tailed rush** (*J. brevicaudatus*) has seeds with only 1 tail, and its flower clusters have non-circular groups of less than 5 flowers, on strongly erect branches. It grows in rock crevices and in fens (rarely) from Newfoundland to Alberta, south to Minnesota, Pennsylvania and North Carolina. **Knotted rush** (*J. nodosus*) has seeds with ridges and no tails, and its flowers form rounded heads. Knotted rush grows on wet soil from Newfoundland to Alaska, south to California, Texas and Virginia. • Two similar rushes in this region lack cross-ridges on their leaves. **Path rush** (*J. tenuis*, p. 149) has leaf sheaths with long, thin, ear-like flaps (auricles) where the sheaths meet the blades, and **Dudley's rush** (*J. dudleyi*, p. 149) has leaf sheaths with short, thick, or no auricles.

Juncus spp., tailed seed

Juncus spp., non-tailed seed

J. brevicaudatus

PATH RUSH
JONC GRÊLE • *Juncus tenuis*

GENERAL: Perennial rush; flowering **stems wiry**, erect, 5–60 cm tall, **tufted**; from fibrous roots.

LEAVES: Mainly basal, flat or inrolled, **less than 1.5 mm wide**, 10–25 cm long, over half as long as flowering stems; sh**eaths have long, white, thin, papery, conspicuous lobes** (auricles) at tip.

FLOWER CLUSTERS: Branching clusters (cymes), 1–8 cm long, slightly crowded to open; **bract slender, leaf-like**, usually overtopping flower cluster; flowers **green**, with 6 pointed, scale-like, 3–4.5 mm long tepals.

FRUITS: Straw-coloured capsules, 3–4 mm long, **dimpled at tip**, shorter than tepals, with many seeds about 0.5 mm long.

WHERE FOUND: In dry and wet situations, often on disturbed sites; from Alaska to Newfoundland, south to Mexico and Florida.

NOTES: **Toad rush** (*J. bufonius*) is a low (usually 5–20 cm tall), annual rush with a soft, branching base and fibrous roots. It has tepals that extend past its capsules, tapering gradually to long slender points, and its capsules have short, pointed tips. Toad rush grows in moist, often sandy, soil on shores, roadsides and disturbed ground from Alaska to Newfoundland, south to California and Florida. • **Dudley's rush** (*J. dudleyi*) is very similar to path rush, but its leaves are shorter (usually less than half the length of the flowering stem), and they have short, firm, yellowish auricles. Also, its flowers are larger, with 4–6 mm long tepals. Dudley's rush grows in dry and wet situations, often on disturbed sites from the southern Yukon to Newfoundland, south to California, Texas, Tennessee and Virginia. • Path rush has been introduced around the world. Its common name refers to its abundance in compacted soil along paths.

J. tenuis

J. bufonius

J. dudleyi

149

SOFT RUSH
JONC ÉPARS • *Juncus effusus*

GENERAL: Perennial rush; flowering stems over 1 m tall, in **dense tussocks** of up to several hundred stems; from stout, branching rhizomes.

LEAVES: Reduced to inconspicuous, **bladeless** sheaths at base of stem.

FLOWER CLUSTERS: Branched panicles of several to many flowers, appearing to arise **from side of stem**; flowers bisexual, with 6 **green or brown, scale-like tepals** 2–2.5 mm long.

FRUITS: Capsules 0.5–2.5 mm long, with 3 to many, tiny, **pale-pointed seeds**.

WHERE FOUND: In marshes and ditches, on streambanks and occassionally in hardwood swamps; from Newfoundland to Alaska, south to California and Florida.

NOTES: Soft rush could be confused with **softstem bulrush** (*Scirpus validus*, p. 144) or **hardstem bulrush** (*S. acutus*, p. 144), both of which also have flower clusters that appear to grow from the side of the stem. However, these bulrushes do not grow in dense tussocks, their flowers do not have 6 tepals, and each of their flowers produces a single achene, rather than a many-seeded capsule. • **Thread rush** (*J. filiformis*) is very similar to soft rush, but the erect bract below each flower cluster is very long, equalling the stem or longer, so that the flowers appear to grow from the middle of the stem or lower. It is slender, only 10–40 cm tall, and close examination of the flowers will show that they have 6 (rather than 3) stamens. Its capsules are 2.4–3.7 mm long. Thread rush grows in fens and along shores from Alaska to Newfoundland, south to Oregon, Utah, Minnesota and West Virginia. • Rushes offer protection for waterfowl, marsh birds and songbirds. The capsules are sometimes eaten by waterfowl, and the rhizomes are occasionally eaten by muskrats and moose. • Native people wove rushes into baskets, bags, mats and clothes. Bunches of dried stems, soaked in animal fat, have been used as torches. • The Ojibway name for soft rush is *kizaebunushkoon*. The genus name *Juncus* means 'binder,' and it refers to the stems, which were used for weaving baskets. The species name *effusus* means 'vast or extensive,' perhaps in reference to the large tussocks formed by this species.

BROWN-FRUITED RUSH
JONC À FRUITS BRUNS • *Juncus pelocarpus*

GENERAL: Slender, perennial rush; flowering stems erect, 5–50 cm tall; from slender, whitish, **spreading rhizomes**; often forms **dense mats**; submerged sterile plants, consisting of flattened rosettes of reddish leaves, sometimes present.

LEAVES: Few, basal and on stem, **thread-like, rounded**, seldom over 13 cm long; sheath loose, with lobes (auricles) at top.

FLOWER CLUSTERS: Loose, spreading clusters **(cymes)**, up to 10 cm long, with small, scattered, **1–2-flowered heads**; flowers **green to reddish green**, often replaced by small, **slender clusters of tiny, firm leaves (bulblets)**, which soon fall; tepals about 2.5 mm long, oblong, usually rounded at tip; 6 stamens, with large anthers that are distinctly longer than their stalks (filaments).

FRUITS: Linear capsules, tapering to a slender point, longer than tepals, with many seeds about 0.5 mm long.

WHERE FOUND: On wet sandy, mucky or muddy shores of lakes, ponds, fens and streams, often with a seasonally variable water level; from Ontario to Newfoundland, south to Minnesota, Indiana and Delaware.

NOTES: Jointed rush (*J. articulatus*) also has a spreading flower cluster above a small bract, but its heads are larger (3–10-flowered), and the stalks (filaments) of its 6 stamens are equal to or longer than their anthers. It grows on sandy or peaty shores and in streams and moist places from Alaska to Newfoundland, south to California, New Mexico, Indiana and North Carolina. • **Bayonet rush** (*J. militaris*) is very similar to brown-fruited rush, but its stems and rhizomes are stouter, its upper stem leaves are bladeless, and its flowers are larger (3–4 mm long) and more numerous (6–12 per head), without bulblets. Its rhizomes produce dense, submerged clusters of thread-like, knobby leaves that can reach 6 m in length. Bayonet rush grows in shallow water, on wet shores and occasionally in fens from Ontario to Newfoundland, south to Michigan and Maryland. It is rare and local in Ontario.

J. militaris

J. militaris

AQUATICS

The aquatic section is divided into plants with submerged, floating and emergent leaf types, with some plants having a combination of these features. Simple, pictorial keys for submerged, floating and emergent plants use leaf and flower characteristics to separate groups of species. Aquatic plants are highly variable in form, but these keys will usually work for the common aquatics.

Polymorphic plants have more than 1 form. The arrowheads (*Sagittaria* spp., pp. 160–61) and pondweeds (*Potamogeton* spp., pp. 171-76) are good examples of plants with leaf dimorphism (2 different types of leaf shape). Distinctive differences in size and leaf characteristics are found in both wetland and terrestrial plants, and they can be attributed to nutrient availability and other environmental factors. Leaf polymorphism is particularly evident in aquatic plants, which can have combinations of 3 types of leaves: submerged, floating and emergent. Factors that influence the variations include age, environmental conditions, nutrient availability, photoperiod, temperature and water level fluctuations.

Some aquatic plants are very difficult to identify because they do not fit into the characteristic description for their species. Identification may be possible with more detailed keys and descriptions, using flower and/or fruit characteristics. With experience one can learn to recognize the different forms of a species growing under various conditions.

KEY TO AQUATIC PLANTS - LEAVES FLOATING

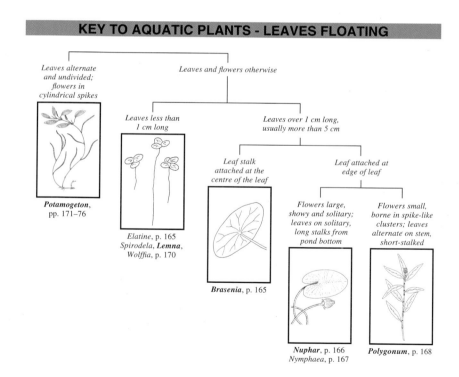

Leaves alternate and undivided; flowers in cylindrical spikes

Potamogeton, pp. 171–76

Leaves and flowers otherwise

Leaves less than 1 cm long

Elatine, p. 165
Spirodela, **Lemna**, Wolffia, p. 170

Leaves over 1 cm long, usually more than 5 cm

Leaf stalk attached at the centre of the leaf

Brasenia, p. 165

Leaf attached at edge of leaf

Flowers large, showy and solitary; leaves on solitary, long stalks from pond bottom

Nuphar, p. 166
Nymphaea, p. 167

Flowers small, borne in spike-like clusters; leaves alternate on stem, short-stalked

Polygonum, p. 168

KEY TO AQUATIC PLANTS - LEAVES EMERGENT

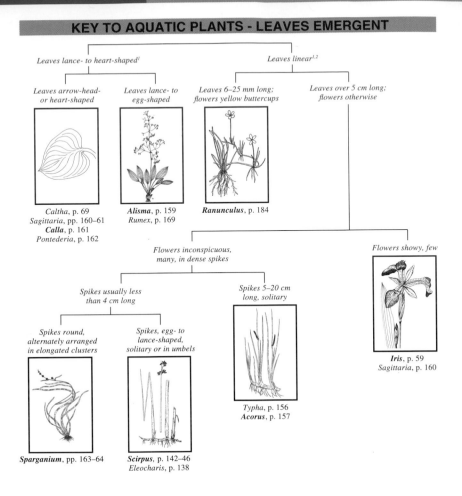

Leaves lance- to heart-shaped[1]

Leaves linear[1,2]

Leaves arrow-head-
or heart-shaped

Leaves lance- to
egg-shaped

Leaves 6–25 mm long;
flowers yellow buttercups

Leaves over 5 cm long;
flowers otherwise

Caltha, p. 69
Sagittaria, pp. 160–61
Calla, p. 161
Pontederia, p. 162

Alisma, p. 159
Rumex, p. 169

Ranunculus, p. 184

Flowers inconspicuous,
many, in dense spikes

Flowers showy, few

Spikes usually less
than 4 cm long

Spikes 5–20 cm
long, solitary

Spikes round,
alternately arranged
in elongated clusters

Spikes, egg- to
lance-shaped,
solitary or in umbels

Iris, p. 59
Sagittaria, p. 160

Sparganium, pp. 163–64

Scirpus, p. 142–46
Eleocharis, p. 138

Typha, p. 156
Acorus, p. 157

---

1. See also key to herbs (p. 50)

2. See also Equisetum fluviatile (p. 194) and keys to grasses (p. 106), and sedges (pp. 116–117).

KEY TO AQUATIC PLANTS - LEAVES SUBMERGED

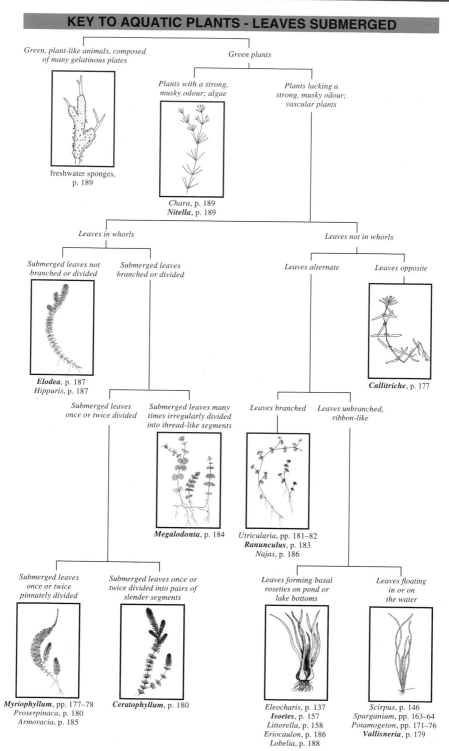

Green, plant-like animals, composed of many gelatinous plates

freshwater sponges, p. 189

Green plants

Plants with a strong, musky odour; algae

Chara, p. 189
Nitella, p. 189

Plants lacking a strong, musky odour; vascular plants

Leaves in whorls

Submerged leaves not branched or divided

Elodea, p. 187
Hippuris, p. 187

Submerged leaves branched or divided

Submerged leaves once or twice divided

Submerged leaves many times irregularly divided into thread-like segments

Megalodonta, p. 184

Submerged leaves once or twice pinnately divided

Myriophyllum, pp. 177–78
Proserpinaca, p. 180
Armoracia, p. 185

Submerged leaves once or twice divided into pairs of slender segments

Ceratophyllum, p. 180

Leaves not in whorls

Leaves alternate

Leaves opposite

Callitriche, p. 177

Leaves branched

Leaves unbranched, ribbon-like

Utricularia, pp. 181–82
Ranunculus, p. 183
Najas, p. 186

Leaves forming basal rosettes on pond or lake bottoms

Eleocharis, p. 137
Isoetes, p. 157
Littorella, p. 158
Eriocaulon, p. 186
Lobelia, p. 188

Leaves floating in or on the water

Scirpus, p. 146
Sparganium, pp. 163–64
Potamogeton, pp. 171–76
Vallisneria, p. 179

155

T. latifolia

COMMON CATTAIL
QUENOUILLE À FEUILLES LARGES • *Typha latifolia*

GENERAL: Perennial herb; stems over 1 m tall; from spreading by rhizomes.

LEAVES: Flat, 10–25 mm wide, sheathing at base, spongy but very strong due to framework of fibres.

FLOWERS: Tiny, in dense spikes, with **male spike at stem tip touching 10–20 cm long, female spike** immediately below; female spikes green in early summer, brown with age.

FRUITS: Minute achenes, with many brown hairs which give mature spikes their brown colour, produced in great quantity.

WHERE FOUND: In marshes, ponds and ditches, and less frequently in fens and swamps; from Newfoundland to Alaska, south to California, Texas and Florida.

NOTES: Narrow-leaved cattail (*T. angustifolia*) has leaves that are only 5–10 mm wide, and its male spike (at the stem tip) is separated from the lower, female spike. Narrow-leaved cattail is found from New Brunswick to southern Manitoba, south to California, Nebraska and South Carolina. Common and narrow-leaved cattails sometimes hybridize. • Established cattail colonies release a substance that is toxic to their own seeds, thus preventing overpopulation. Common cattail quickly colonizes new and disturbed habitats, using its fluffy masses of wind-dispersed seeds and its spreading rhizomes. • Cattails are a good source of nest-building material. Birds, such as marsh wrens, pied-billed grebes, American coots and sora, use cattails to lash their nests together. Other birds that nest in the cattails include red-winged blackbirds, swamp sparrows and American bitterns. Canada geese eat the rhizomes and occasionally the young spikes. • If you have ever seen a mound of cattail leaves on the side of pond or marsh you have located the home of a muskrat. Its main food, whether for convenience or taste, is the cattail rhizome. • The bulky winter spikes are the overwintering home and food supply of the cattail moth. This moth lays its eggs in the loose downy seeds, which it then binds together with silk. • In Ojibway, common cattail is called *pukawayaushkawi* (the common reed that splits), because the leaves were split in 2 and used for weaving. • To the native people of North America, this plant was like the local supermarket: they made flour from the pollen, and they dried rhizomes; they ate the young stems like asparagus; they cooked the green spikes like corn on the cob (the taste is identical); and they cooked the rhizomes (which contain over 30% starch and sugar) like potatoes. Common cattail can be used to make a powder similar to corn starch, and it is said that an excellent syrup can be extracted from the roots. • Cattails also served other needs of native people. The seed fluff was used as a warm down filling for bedding and clothes, and warriors reportedly threw it into the eyes of their enemies to cause blindness.

T. angustifolia   *T. angustifolia*

SWEETFLAG, CALAMUS
ACORUS ROUSEAU • *Acorus calamus*

GENERAL: Perennial, emergent herb, **very aromatic**; stems flattened, resembling leaves, up to 2 m tall; forming loose colonies from thick rhizomes.

LEAVES: Long, stiff, basal, 8–25 mm wide, **4-sided (diamond-shaped in cross-section)**; midveins not in centre of leaf.

FLOWERS: Minute, yellow-brown, in a spike-like cluster (spadix) that is 5–10 cm long and up to 2 cm thick, appearing to be borne mid-way on stiff 3-sided stalk; mid-summer.

FRUITS: Tiny berries, covering a fleshy spike.

WHERE FOUND: In marshes and along lakeshores and slow streams, usually in shallow water; from Nova Scotia to British Columbia, south to Washington, Texas and Florida.

NOTES: This species is also known as *A. americanus*. • The rhizomes contain strong, volatile oils, which give this plant its distinct aroma. • Sweetflag could be confused with **cattails** (*Typha* **spp.**, p. 156) or **burreeds** (*Sparganium* **spp.**, pp. 163–64), but it can be distinguished by its spicy aroma, off-centre midveins and narrow flower clusters that arise from the side (rather than the tip) of the stem. • Muskrats eat the rhizomes. • Native people called sweetflag *powemenarctic* or 'muskrat root,' and they used it to treat cholera, coughs, toothaches and other ailments. • The species name *calamus* means 'beauty,' which suits this plant very well.

SPINY-SPORED QUILLWORT
ISOÈTE À SOR • *Isoetes echinospora*

GENERAL: Onion-like, evergreen, aquatic, perennial quillwort; stems short, fleshy corm-like; from forked, fibrous roots.

LEAVES: Linear, 5–15 cm long, 0.5–1.5 mm wide, with 4 large, longitudinal air cavities, spoon-shaped and clasping at base, forming a basal rosette.

SPORE CLUSTERS: Solitary, on inner side of spoon-shaped leaf bases, covered by a thin membrane (velum); spores all 1 of 2 types, produced in an alternating cycle; megaspores white, resembling salt grains, 0.3–0.5 mm wide, **with tiny spines (visible with hand lens)**; microspores fawn-coloured, 23–30 microns long, not visible to naked eye; late summer.

WHERE FOUND: Usually submerged on sandy or rocky lakeshores or just above the waterline on wet, sandy shores; from Newfoundland to Alaska, south to California, Indiana and New Jersey.

NOTES: Spiny-spored quillwort can be confused with vegetative forms of **water lobelia** (*Lobelia dortmanna*, p. 188) or **needle spikerush** (*Eleocharis acicularis*, p. 137), but look for its spoon-shaped leaf bases, which house the spore cases. • The megaspores must be closely examined to identify quillworts to species. **Lake quillwort** (*I. lacustris*) has larger (0.5–0.8 mm wide) megaspores that are ridged (rarely smooth) rather than spiny. It is found from Quebec to British Columbia, south to California and Colorado. • The genus name *Isoetes* means 'green throughout the year,' in reference to the evergreen leaves.

SHORE PLANTAIN, SHOREGRASS
LITTORELLE D'AMÉRIQUE • *Littorella americana*

GENERAL: Low, perennial, **fleshy herb**; flowering stems usually shorter than leaves; tufted, **often forming mats** from creeping stems (stolons).

LEAVES: Bright green, **basal, linear, cylindrical**, 27 cm long.

FLOWERS: Urn-shaped, with 4 spreading lobes, whitish or pale, **male or female** on separate plants; male flowers with slender, **conspicuous stamens** projecting from their centres; **female flowers** very small, **usually 2**.

FRUITS: Blackish achenes, about 2 mm long.

WHERE FOUND: Submerged in shallow water of gravel-, sand- or clay-bottomed lakes, or stranded above the water level on beaches; from Minnesota to Newfoundland, south to New York and Maine.

NOTES: This species is also known as *L. uniflora*. It is often in association with **quillworts** (*Isoetes* **spp.**, p. 157), **pipewort** (*Eriocaulon aquaticum*, p. 186), **water lobelia** (*Lobelia dortmanna*, p. 188) and **mud sedge** (*Carex limosa*, p. 128). • If flowers and fruits are lacking, it is difficult to distinguish this species from creeping spearwort (*Ranunculus reptans*, p. 184). Creeping spearwort has solitary yellow flowers at its nodes and shorter leaves (less than 5 cm) in clumps of 24. It also has prostrate, thread-like stems, and it forms dense mats. • **Awlwort** (*Subularia aquatica*) is similar to shore plantain, but it lacks stolons. It has tiny flowers with 4 petals and 6 stamens. Awlwort is found in localized populations from Alaska to Newfoundland, south to California, Montana, central Saskatchewan and New England. Like shore plantain, awlwort is found on sandy shores in shallow water, or stranded above the waterline. Awlwort is in the mustard family (Brassicaceae). • The genus name *Littorella* refers to the shoreline habitat of shore plantain.

L. americana

Subularia aquatica

Subularia aquatica

WATER PLANTAIN
ALISMA PLANTAIN-D'EAU • *Alisma plantago-aquatica*

GENERAL: Perennial, aquatic, herb, with a 30–100 cm tall, leafless flowering stalk; from a fleshy, corm-like base with fibrous roots.

LEAVES: Basal, up to 18 cm long, oval or elliptic, **long-stalked, with parallel veins**; submerged, ribbon-like leaves sometimes present.

FLOWERS: Up to 8 mm wide, white to pink, **3-parted,** with 3 sepals and 3 petals, on 1–4 cm long stalks, in **large open clusters (panicles)**; throughout summer.

FRUITS: Achenes, 2–2.5 cm long, grooved on 1 side and ridged on other side.

WHERE FOUND: In marshes, lakes, streams and ditches; quickly colonizes ditches and other disturbed sites; from Newfoundland to British Columbia, south to California, Florida and Texas.

NOTES: This species is also known as *A. triviale.* • **Narrow-leaved water plantain (*A. gramineum*,** previously known as *A. geyeri*) is a smaller plant, with leaves that are linear if submerged and narrowly lance-elliptic if emergent. Its flower clusters are less conspicuous, as they are usually shorter than, and therefore obscured by, the leaves. The achenes of narrow-leaved water plantain have 1 ridge and 2 grooves (rather than 1 groove). This plant grows in the mud or shallow water of marshes, ponds, lakes and slow-moving streams from southern British Columbia to southern Quebec, south to California, Colorado and New York. • Water plantain achenes are occasionally eaten by waterfowl, but they are not an important food source. Moose sometimes eat the leaves. • The species name *plantago-aquatica* means 'plantain-leaved water plant,' in reference to the similarity of the leaf shape and venation of this plant to that of a plantain (*Plantago* sp.).

A. gramineum   A. plantago-aquatica

STIFF ARROWHEAD
SAGITTAIRE DRESSÉE • *Sagittaria rigida*

GENERAL: Tall, emergent, perennial herb, 10–90 cm tall; from rhizomes and tubers.

LEAVES: Variable, mostly **lance-shaped or linear**, rarely arrowhead-shaped; submerged leaves (when present) up to 50 cm long, thin; in a basal rosette.

FLOWERS: With 3 white petals and **3 green sepals**; in few whorls of 3; **unisexual with male flowers above female flowers on same plant**; throughout summer.

FRUITS: Achenes, stalkless, crowded in rounded heads; mid- to late summer.

WHERE FOUND: In marshes, lakes, ponds, rivers, streams, ditches and swamps; from southern Quebec to northwestern Ontario, south to Nebraska, Tennessee and Virginia.

NOTES: The leaf shape of stiff arrowhead is highly variable, due to changes in water level and other environmental factors. • **Grass-leaved arrowhead** (*S. graminea*) is similiar to stiff arrowhead, but its fruiting heads are stalked. It grows in shallow water and on muddy shores from Newfoundland to Lake Superior, south to Texas, Ohio Florida. • The seeds and tubers (especially the tubers), are valuable food for waterfowl and marsh birds, such as canvasbacks, mallards, scaups, teals, wood ducks, geese and rails. Muskrats and porcupines feed on the leaves and tubers.

FLOATING ARROWHEAD
SAGITTAIRE CUNÉAIRE • *Sagittaria cuneata*

GENERAL: Floating-leaved or emergent, perennial, aquatic herb; emergent form 10–60 cm tall; from rhizomes and tubers.

LEAVES: Typically floating, arrowhead-shaped, with stalks up to 90 cm long; emergent leaves only when plant in shallow water; submerged leaves thin, ribbon-like; in a basal rosette.

FLOWERS: White, showy, 2–3 cm wide, with **bract over 1 cm long at base of flower stalk**; in whorls of 3.

FRUITS: Achenes, tipped with an **erect beak**, in dense, spherical heads.

WHERE FOUND: In quiet, shallow water of lakes and streams; from Nova Scotia to Alaska, south to California, Texas, Iowa and New York.

NOTES: The genus name *Sagittaria* means 'arrow-shaped,' and it refers to the shape of the floating leaves.

BROAD-LEAVED ARROWHEAD, WAPATO
SAGITTAIRE LATIFOLIÉE • *Sagittaria latifolia*

GENERAL: Perennial, emergent herb, 20–80 cm tall; from rhizomes and tubers.

LEAVES: Emergent leaves variable, mostly **arrowhead-shaped**, up to 40 cm long; submerged leaves (when present) **thin and ribbon-like**, in a **basal rosette.**

FLOWERS: White, showy, 2–4 cm wide, with **3 white petals, 3 green sepals** and a bract less than 1 cm long, **unisexual, with male flowers above female flowers on same plant**, in 2–8 whorls of 3; throughout summer.

FRUITS: Achenes, with **beaks projecting horizontally from tips**, crowded in rounded heads; mid- to late summer.

WHERE FOUND: In marshes, lakes, ponds, streams and wet ditches; from Quebec to British Columbia, south to California and Texas.

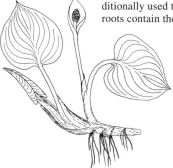

NOTES: The leaf shape of broad-leaved arrowhead is highly variable, due to changes in water level and other environmental factors. • The seeds and tubers (especially the tubers) are valuable food for waterfowl and marsh birds, such as canvasbacks, mallards, scaups, teals, wood ducks, geese and rails. Muskrats and porcupines eat the leaves and tubers. • The Ojibway and other native tribes used the tubers for food. They dug them from the mud or collected them from muskrat lodges.

WATER ARUM, WILD CALLA
CALLA DES MARAIS • *Calla palustris*

GENERAL: Perennial, aquatic herb, 10–30 cm tall; from **long, creeping rhizomes**.

LEAVES: Basal, heart-shaped, long-stalked; blades 5–10 cm long.

FLOWERS: Tiny, lacking petals, in a dense, pinkish cluster on a **fleshy spike (spadix)** immediately above and hooded by a **white**, 1.5–2.5 cm long, **oval bract (spathe)**; early to mid-summer.

FRUITS: Red, fleshy, few-seeded berries, in dense heads; mid- to late summer.

WHERE FOUND: In swamps, marshes and ditches and on streambanks; from Newfoundland to Alaska, south to Colorado, Texas and Florida.

NOTES: The dried seeds and rhizomes were traditionally used to make flour. The fresh plant and roots contain the **poison** calcium oxalate, but this compound is destroyed in the flour making process. Calcium oxalate can cause intense burning, irritation and swelling of the mouth, tongue and lips. This is the same poisonous compound found in rhubarb. • Muskrats occasionally eat water arum. • An Algonquin Indian tribe calls this plant *ehawshoga*, which means 'plant that bites back at the mouth.' Native people used this plant to treat sore eyes, colds, fevers, swellings and snake bites.

PICKERELWEED
PONTÉDÉRIE CORDÉE • *Pontederia cordata*

GENERAL: Perennial, **aquatic herb**; stems erect, stout, 30–60 cm tall, bearing a single leaf; from thick, spreading rhizomes.

LEAVES: Mainly **basal**, thick, with many parallel veins, **lance- to egg-shaped, heart-shaped** at base and blunt-tipped, 5–25 cm long, 2–15 cm wide, with long, sheathing stalks.

FLOWERS: Violet-blue with 2 yellow dots on upper lip, funnel-like, about 8 mm long, 2-lipped with a 3-lobed upper tepal and 3 spreading lower tepals; in a **dense, glandular-hairy, stalked spike** above a large bract; soon fading, July–September.

FRUITS: 1-seeded bladders (urticles), enclosed in rough-ribbed, persistent bases of flower tubes.

WHERE FOUND: In shallow water (rarely more than 1 m deep) along muddy or sandy shores of lakes, ponds, marshes, rivers and streams; often forming large, dense colonies in still waters of coves and bays; from Ontario to Nova Scotia, south to Texas and Florida.

NOTES: Pickerelweed could be mistaken for non-flowering **arrowheads** (*Sagittaria* spp., pp. 160–61) that have narrow leaves, but the leaf stalks of pickerelweed have a basal sheath, the leaf blades are usually heart-shaped, with more veins, and the plants lack tubers. • **Water plantain** (*Alisma plantago-aquatica*, p. 159) has leaves with prominent midveins and flowers arranged in branched clusters (panicles). • **Mud plantain** (*Heteranthera dubia*, also known as *Zosterella dubia*), also called **water star-grass**, is a much smaller, less conspicuous member of the pickerelweed family. It is a slender, submerged plant with grass-like leaves less than 1 cm wide, which broaden to the base (but are not heart-shaped) and are borne on slender, branching stems. The small, pale yellow flowers of mud plantain arise singly at the tips of long, thread-like tubes that carry the flower to the water surface. This aquatic plant is less common than pickerelweed, and it grows on muddy shores, in streams and in the quiet waters of bays and lakes from Washington to Quebec, south through Mexico to tropical America. The leaves of mud plantain vary greatly in width, and this plant is often confused with a few of linear-leaved **pondweeds** (*Potamogeton* spp., pp. 171–76). Look for an indistinct midvein in the leaf of mud plantain, compared to the distinct midveins of pondweed leaves. • The young stems and leaves of pickerelweed can be eaten in salads or boiled and served with butter. The seeds are also eaten raw. • Muskrats and waterfowl eat pickerel-weed seeds, and deer and muskrats sometimes eat the leaves. • This plant is pollinated by butterflies, bees and flies. • Pickerelweed is known as *kinozhaeguhnsh* in Ojibway, which means 'pike's plant.' This genus was named in honour of Giulio Pontedara (1688–1757), professor of botany in Padua, Italy. The species name *cordata* refers to the cordate (heart-shaped) leaves.

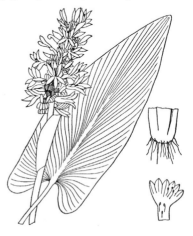

LARGE-FRUITED BURREED
RUBANIER À GROS FRUITS • *Sparganium eurycarpum*

GENERAL: Robust, emergent, perennial herb, up to 1.2 m tall; stems stout, with a **characteristic zigzag pattern**; from fibrous roots and creeping rhizomes.

LEAVES: Alternate, narrow, stiff, 2-ranked, **keeled (V-shaped)**, spongy at base; flexible, ribbon-like, submerged leaves up to 80 cm long develop before stem.

FLOWERS: Tiny; unisexual, with spherical, greenish, 1–2 cm wide, male heads above **bur-like**, 3–4 cm wide female heads; mid- to late summer.

FRUITS: Nut-like achenes, flat at tip but with a beak-like, **2-pronged, persistent style**; late summer.

WHERE FOUND: In shallow (less than 1 m deep) water in quiet marshes, rivers and ponds; from Newfoundland to British Columbia, south to California, Kansas and Ohio.

NOTES: Two common burreeds in this area have 1-pronged styles and erect leaves. **Common burreed** (*S. emersum*, formerly known as *S. chlorocarpum*) has flat, 6–12 mm wide leaves, and it is smaller than large-fruited burreed (usually less than 30 cm tall). Its long-stalked fruits are greenish at the base, and they have beaks more than 2 mm long. Common burreed is found from Newfoundland to southern Alberta, south to California, South Dakota and North Carolina. **American burreed** (*S. americanum*) has 2 or more male flowerheads, and its fruiting heads are more than 12 mm in diameter, with beaks longer than 1.5 mm. American burreed is found from Newfoundland to to British Columbia, south to Texas and Florida. • Several species of aquatic plants may be confused with submerged leaves of burreed. These include **tape grass** (*Vallisneria americana*, p. 179), **wild rice** (*Zizania palustris*, p. 115), **arrowheads** (*Sagittaria* **spp.**, pp. 160–61) and **water plantain** (*Alisma plantago-aquatica*, p. 159). The leaves of burreed can be distinguished by their many veins and by the checkered pattern caused by cross veins between the parallel veins. Cross veins can also be found in the submerged leaves of arrowhead, but they are not as numerous. • The emergent burreed leaves may resemble the leaves of **cattail** (*Typha* sp., p. 156) or **sweetflag** (*Acorus calamus*, p. 157), but burred leaves are triangular in cross-section. • Many birds, including mallards, coots, black ducks, wood ducks, common snipes and rails, eat large-fruited burreed fruits. The stems and leaves are a preferred food of muskrats and deer.

S. emersum

S. eurycarpum

S. emersum

163

FLOATING-LEAVED BURREED
RUBANIER FLOTTANT • *Sparganium fluctuans*

GENERAL: Perennial, aquatic herb; stems floating; from fibrous roots and creeping rhizomes.

LEAVES: Elongated, flat, 3–10 mm wide, 20–100 cm long, **translucent at edges, floating.**

FLOWERS: Unisexual, in rounded male or female heads on same plant; male heads above bur-like, **1.5–2 cm thick,** female heads.

FRUITS: Nut-like achenes, 3–4 mm long, tipped with a **single, beak-like,** persistent, curved style; mid- to late summer.

WHERE FOUND: In marshes, rivers and ponds; from Newfoundland to northern Alberta, south to Minnesota, Pennsylvania and New England.

NOTES: There are other species of burreeds with floating leaves and single styles. • **Least burreed** (*S. natans*, formerly known as *S. minimum*) has 1 male flowerhead, and its fruiting heads are less than 12 mm in diameter, with beaks shorter than 1.5 mm. It is found from Newfoundland to Alaska, south to Oregon, Minnesota and New Jersey. • **Northern burreed** (*S. hyperboreum*) has 1 male flowerhead, flat leaves and dark yellow fruits with tiny (less than 0.5 mm long) beaks. It is found from Newfoundland to Alaska, south to southern Manitoba and Nova Scotia. • **Narrow-leaved burreed** (*S. angustifolium*) resembles floating-leaved burreed, but its leaves are 2–5 mm wide, and they are rounded on the back rather than flat. Its fruiting heads are 1–2 cm in diameter, and the beaks of its achenes are slender, greater than 1.5 mm long and straight, rather than hooked to 1 side. Narrow-leaved burreed grows in the shallow water of lakes, ponds and slow-moving streams from Alaska to Newfoundland, south to California, Colorado, Minnesota and New Jersey. • Floating-leaved burreed fruits are eaten by many birds, including mallards, coots, black ducks, wood ducks, common snipes and rails. The stems and leaves are a preferred food of muskrats and deer. • Burreed tubers have been collected and eaten like potatoes, but they are very difficult to gather because they are small and scattered. These plants should be gathered **cautiously** and in moderation, due to the sensitivity of wetland environments, and because of possible confusion with **toxic tubers** of poisonous wetland plants.

S. angustifolium

LONG-STEMMED WATERWORT
ÉLATINE À TROIS ÉTAMINES • *Elatine triandra*

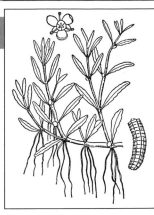

GENERAL: Limp, tufted, hairless, **annual** herb; **stems** 5–10 cm long, **creeping**, branched and **rooting at joints**.

LEAVES: Thin, lance-shaped, often widest above middle, 2–8 mm long, stalkless.

FLOWERS: Minute, **single in leaf axils**, stalkless, 3-parted.

FRUITS: Membraneous capsules, usually 3-parted, with several to many slender, curved seeds (less than 0.8 mm long) attached at different levels in fruit.

WHERE FOUND: On mud or in shallow water along lake, pond and river edges; from the Northwest Territories to New Brunswick, south to Mexico, Missouri and Virginia.

NOTES: Small waterwort (*E. minima*) is smaller than long-stemmed waterwort, with leaves rarely over 5 mm long and 3 mm wide, and its capsules and flowers are usually 2-parted (rather than 3-parted). The general form of waterworts is influenced by water level, and these 2 species can only be positively identified by position of the seeds in their fruits. Small waterwort seeds are all borne at the same level in the fruit. This species grows submerged in shallow water on sandy, clayey or silty shores (often with **awlwort** [*Subularia aquatica*, p. 158], **mud sedge** [*Carex limosa*, p. 128] and **pipewort** [*Eriocaulon aquaticum*, p. 186]) or emergent on sandy shores from Minnesota to Newfoundland, south to Wisconsin and Virginia. Both of these species are considered rare in Ontario, and they are inconspicuous and easily overlooked. Also, they could be mistaken for seedlings of **St. John's-wort** (*Hypericum* sp., p. 78), if the fruit is absent. • Ducks eat waterworts. • The species name *triandra* means '3 stamens,' in reference to the 3-parted flowers of long-stemmed waterwort. The species name *minima* means 'small,' alluding to the leaves of small waterwort.

WATER SHIELD
BRASÉNIE DE SCHREBER • *Brasenia schreberi*

GENERAL: Perennial, aquatic herb; **submerged stem covered with a thick, gelatinous film**; from creeping rhizomes.

LEAVES: Oval, not split, 4–12 cm long, coated with slime beneath and with a **long slimy stalk attached at centre**; mature **leaves floating**.

FLOWERS: Dull, purple-red, 3-parted with 3, 12–16 mm long petals and 3 similar sepals, raised slightly above surface of water on thick stalks; early summer.

FRUITS: Leathery, club-shaped pods, not opening, containing **1–2 seeds**; mid- to late summer.

WHERE FOUND: In shallow, quiet lakes, ponds and streams; from Nova Scotia to Alaska, south to California, Texas and Florida.

NOTES: Where it is abundant, the hard pods (carpels) are often eaten by waterfowl, such as mallards, pintails, redheads and wood ducks. Moose feed on the leaves. • Young water shield plants are said to make a delicious, though slimy, salad, and the roots have traditionally been made into flour.

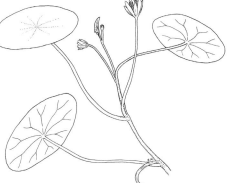

YELLOW POND LILY
NÉNUPHAR À FLEURS PANACHÉES • *Nuphar variegatum*

GENERAL: Perennial, **aquatic herb**; from scaly, thick, yellowish rhizomes.

LEAVES: Floating, pinnately veined, **heart-shaped** with rounded lobes, **10–25 cm long**, submerged when young, floating when mature; **stalks** strongly flattened, **winged**, up to 2 m long.

FLOWERS: Showy, yellow, 4–6 cm wide, floating on or slightly raised above water surface, usually with **6 showy, yellow, petal-like sepals and many small, scale-like, yellow petals merging with yellow anthers**; stigma disc green or yellowish; throughout summer.

FRUITS: Leathery berries, maroon or red, oval, 2–4.5 cm long, with many seeds; mid- to late summer.

WHERE FOUND: In lakes, ponds, quiet streams and rivers; from Newfoundland to the Yukon Territory, south to Idaho, Ohio and Delaware.

NOTES: Small yellow pond lily (*N. pumila*, also known as *N. microphyllum*) is uncommon in northern Ontario. It is smaller than yellow pond lily, and it has only 5 showy sepals and a red stigma disc. Small yellow pond lily is found from Newfoundland to Manitoba, south to Minnesota, Pennsylvania and New Jersey. • Waterfowl and marsh birds, such as black ducks, mallards, pintails, ringneck ducks, teals, wood ducks, bitterns and soras, eat the seeds. Beavers, muskrats, porcupines and deer eat the rhizomes and leaves. Yellow pond lily is a favourite food for moose, which feed on the stems and rhizomes. The fat, scaly rhizomes float to the surface when they are uprooted. The leaves provide food and shelter for fish and invertebrates. • Yellow pond lily seeds can be fried into 'popcorn' or cooked like sweet corn, and the large rhizomes can be used like potatoes. Native people gathered the rhizomes from muskrat houses, and then they cooked them like potatoes. The rhizomes were also used to heal cuts and swelling.

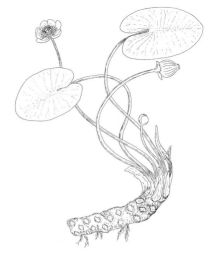

FRAGRANT WHITE WATER LILY
NYMPHÉA ODORANT • *Nymphaea odorata*

GENERAL: Perennial, aquatic herb; from thick, scaly, horizontal rhizomes.

LEAVES: Floating, palmately veined, 7–30 cm long, rounded, with narrow V-shaped split; stalks long, rounded, cross-section reveals 4 large air passages used to pump oxygen to roots.

FLOWERS: White, showy, fragrant, 7–20 cm wide, with many white petals and 4 green sepals, on a rounded receptacle, floating on surface of water; open only from mid-morning to early afternoon; throughout summer.

FRUITS: Leathery berries, with many seeds; ripening under water in mid- to late summer.

WHERE FOUND: In lakes, ponds, slow rivers and marshes; from Newfoundland to Manitoba, south to Texas and Florida.

NOTES: **Northern small white water lily** (*N. tetragona*) has smaller, 3–8 cm long, non-fragrant flowers on 4-sided receptacles, it has smaller, 4–12 cm wide, rounded leaves with a wide split, and its flowers open in the afternoon. It grows in cold streams and lakes, from Quebec to Alaska, south to Minnesota and Maine. • **Floating heart** (*Nymphoides cordata*, previously known as *Nymphoides lacunosum*) with its 1.5–6 cm wide, rounded or heart-shaped, floating leaves could be mistaken for a tiny water lily, but it is more closely related to **buckbean** (*Menyanthes trifoliata*, p. 91).

The floating leaves are anchored to their rhizomes by long, slender stalks (less than 1 mm thick near the top), and just below the leaf blade, they produce unbranched clusters (umbels) of white, 5–12 mm wide, 5-petalled flowers, often with a dense cluster of short, thick roots. Floating heart grows in shallow, quiet water of ponds and lakes from Ontario to Newfoundland, south to Louisiana and Florida. • Waterfowl, such as wood ducks, teals and scaups, eat the fruits, seeds and rhizomes of fragrant white water lily. The plants (especially the rhizomes) are eaten by muskrats, beavers, deer and porcupines. Moose feed on the leaves, but they prefer yellow pond lily. The leaves also provide shade and habitat for fish and invertebrates. • The Ojibway ate fragrant white water lily flower buds. They also wrapped food in the leaves for cooking. They made a tea from the rhizomes to treat diarrhea, sore throats and balding, and they used powdered dried rhizomes to treat boils, ulcers and sores and to make cough medicine.

N. tetragona

WATER SMARTWEED
RENOUÉE AMPHIBIE • *Polygonum amphibium*

GENERAL: Perennial, aquatic herb, with a floating aquatic form and an erect terrestrial form, up to 1 m tall; **stems with swollen joints surrounded by papery sheaths (ocreas)**, rooting at submerged nodes; from rhizomes.

LEAVES: Alternate, short-stalked or stalkless, with leaf-like bracts at base forming a sheath (ocrea); aquatic leaves broadly oval, 2–15 cm long, blunt, floating, **often reddish**; terrestrial leaves 6–12 cm long, lance-shaped, pointed.

FLOWERS: 4–5 mm wide, in dense, elongated clusters at stem tips (spike-like racemes), usually **bright pink**, 5-parted, with 5 tepals; mid- to late summer.

FRUITS: Dark, 2-sided achenes; late summer to early autumn.

WHERE FOUND: In shallow ponds, lakes, rivers, streams and marshes and on wet shorelines; from Newfoundland to Alaska, south to Utah, Nebraska and New Jersey.

NOTES: The genus name *Polygonum* means 'many swollen joints'; this is an important identifying feature. • **Lady's thumb** (*P. persicaria*, p. 67) has pink flowers (often in erect, narrow clusters), erect stems, taproots, ocreas that are fringed with long hairs and leaves with a dark patch in the centre. • **Arrow-leaved smartweed** (*P. sagittatum*) has white flowers in loose clusters, arrowhead-shaped leaves, and stems with prickles. It grows in marshes and wet meadows from Newfoundland to southern Manitoba, south to Texas and Florida. • **Pale smartweed** (*P. lapathifolium*) has pink, white or green flowers in nodding, narrow clusters (spikes), long, narrow leaves, and ocreas that are not fringed. It grows on shorelines and in damp to swampy areas from Newfoundland to British Columbia, south to California, Minnesota and New Jersey. • **Marsh pepper smartweed** (*P. hydropiper*) has green flowers in nodding spikes and long, lance-shaped leaves. It grows on damp soil and in ditches from Alaska to Newfoundland, south to California, Texas and Alabama. • **Pennsylvania smartweed** (*P. pensylvanicum*) is a large, annual smartweed, 30–90 cm tall, with lance-shaped leaves that are 5–25 cm long. Its leaves are hairless on their lower surface and on their sheaths (ocreae), and its pink or white flower spikes are borne in erect branching clusters (racemes). Pennsylvania smartweed grows on moist shores, river edges, marshes and mucky hollows from Ontario to Nova Scotia, south to Texas and Florida. • Many birds, including black ducks, mallards, pintails, teals, wood ducks, rails, Canada geese, sparrows and redpolls, feed heavily on smartweed seeds, which are a winter staple for some of them. Mammals, such as chipmunks, squirrels, mice and deer, eat the plants and fruits. • North American native people used water smartweed to cure stomach pains (ulcers), headaches and hemorrhoids, and as an antiseptic for cleansing and healing wounds. Water smartweed has a hot peppery taste. It has been used in salads, and it can be cooked like spinach. The fresh leaves are very high in vitamins K and C, and the high level of vitamin K (a coagulating substance) is probably why native people found it useful for healing wounds.

GREAT WATER DOCK
RUMEX ORBICULAIRE • *Rumex orbiculatus*

GENERAL: Perennial herb, up to 2.5 m tall; stem unbranched; from stout taproot.

LEAVES: Simple, alternate, long-stalked, leathery, flat, **up to 60 cm long**.

FLOWERS: 5–8 mm long; green, turning brown, in branched flower clusters (panicles) up to 50 cm long; mid-summer.

FRUITS: Brown, sharply triangular achenes, enclosed in heart-shaped sepals (valves); late summer.

WHERE FOUND: On open shores and in marshes, wet meadows and floating mat fens; sometimes rooted in shallow water. Newfoundland to North Dakota, south to Indiana and Nebraska.

NOTES: Great water dock was formerly known as *R. britannica*. • **Western dock** (*R. occidentalis*) has leaves with heart-shaped bases and wavy edges. It grows on damp soil from Alaska to Newfoundland, south to California, Texas and Maine. • **Golden dock** (*R. maritimus*) has lance-shaped leaves with wavy edges, and its branching stem is up to 60 cm tall. Golden dock grows in marshes and on shores from Alaska to Nova Scotia, south to California, Arkansas and New York. • **Water dock** (*R. verticillatus*) has leaves in distinct whorls. It grows in swampy areas from Minnesota to Quebec, south to Texas and Florida. • **Willow-leaved dock** (*R. triangulivalvis*, previously known as *R. mexicanus*) has flat, lance-shaped leaves. It grows in open, moist sites from Alaska to Newfoundland, south to Mexico, Ohio and New England. • Several other introduced species of dock are found in disturbed, upland areas. • Docks have been used as cooked greens, in salads and as an antiseptic to treat skin conditions. They should be used in moderation, however, because they contain high concentrations of tannins and oxalic acid which can cause kidney problems.

R. orbiculatus

R. occidentalis

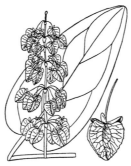

R. occidentalis

R. verticillatus

169

L. minor (larger plants),
W. arrhiza (smaller plants)

S. polyrhiza

L. minor

L. trisulca

S. polyrhiza

DUCKWEEDS AND WATERMEALS
Lemna spp., *Spirodela* spp., *Wolffia* spp.

GENERAL: Small, free-floating, leaf-like thallus (frond), not differentiated into a leaf and stem, with 1 to many **unbranched roots** hanging below; commonly produce vegetative buds.

LEAVES/PLANTS: Small, green, flattened, rounded to oval, obscurely nerved, solitary or (more commonly) in large colonies.

FLOWERS: Rare, in 2 pouches on edge near base of frond, male or female, reduced to a single, flask-shaped ovary or 1–2 stamens; mid-summer.

FRUITS: Tiny, thin-walled utricles, with 1–few seeds; mid- to late summer.

WHERE FOUND: Usually in quiet ponds, stagnant rivers, pools in swamps and standing water in ditches.

NOTES: Duckweeds and watermeals are among the smallest plants in the world. Three species of duckweed are fairly common in northern Ontario. A 10x hand lens will help you to see the details of these tiny plants. • **Lesser duckweed** (*L. minor*) has stalkless fronds with 1 root and 1–3 nerves. Its plants are less than 6 mm across, green above and below, and they float singly or in small clusters on the water surface. It is found from Newfoundland to Alaska, south to California, Texas and Florida. • **Star duckweed** (*L. trisulca*) has branched, stalked fronds with 1 root and 1–3 nerves. Its plants are 6–15 mm long, and they float at or just below the water surface. Star duckweed is gound from Nova Scotia to Alaska, south to California, Texas and Florida. • **Greater duckweed** (*S. polyrhiza*) has fronds with 2 or more roots and 3–11 nerves. They are 5–8 mm across, and their undersides are solid reddish purple, rather than green like those if the *Lemna* species. Greater duckweed is found from Nova Scotia to southern British Columbia, south to California, Texas and Florida • **Columbia watermeal** (*W. arrhiza*, also known as *W. columbiana*) and **dotted watermeal** (*W. borealis*, also known as *W. punctata*) are tiny, rounded, floating plants that could be mistaken for duckweed. However, they are even smaller than the duckweeds (only 0.5–1.5 mm long), and they lack roots altogether. Columbia watermeal is identified by its smooth (not dotted), strongly convex upper surface, whereas dotted watermeal has a flat upper surface that is covered with minute brown dots. The upper surface of Columbia watermeal floats just above the water, but dotted watermeal floats just below the surface. Both watermeals grow in still waters from Minnesota to Ontario, south to Texas and Florida, with occasional reports from Quebec and the prairie provinces. • Small, floating liverworts, such as **purple-fringed liverwort** (*Ricciocarpos natans*, p. 205), could be mistaken for large duckweeds. • Waterfowl and marsh birds, such as coots, black ducks, mallards, teals, wood ducks, buffleheads and rails, eat duckweed in great numbers. Muskrats and beavers occasionally eat duckweeds. These plants also provide food and shelter for many aquatic invertebrates and fish.

Pondweeds • *Potamogeton* spp.

Pondweeds have stems that are long and flexible, floating in or on the water, and they are usually anchored by roots and rhizomes. The leaves can either be long, narrow and submerged and/or floating and oval-shaped. Pondweeds are considered a very difficult group of plants to identify. The seeds are often needed to confirm identifications, but they do not fruit until late into the season. Also, their leaves are highly variable (depending on water conditions), and the different species hybridize. Fifteen of the most abundant pondweeds in Ontario are described in this guide, and several of these are very distinctive.

some leaves broad and floating (pp. 171–73)	*leaves all lance-shaped* (p. 174)	*leaves all linear* (pp. 175–76)

LARGE-LEAVED PONDWEED
POTAMOT À GRANDES FEUILLES • *Potamogeton amplifolius*

GENERAL: Large, robust, submerged perennial herb; stems rounded, 2–4 mm thick, unbranched or with a few branches above; from rhizomes.

LEAVES: Alternate, stalked, parallel-veined, wavy-edged, rounded and not clasping at base, with **leaf-like bracts (stipules)**; **submerged leaves** brownish, **3–7 cm wide, 8–20 cm long; floating leaves usually present, elliptic, leathery, waxy on upper surface**.

FLOWERS: 4–5.5 mm wide, with 4 sepal-like bracts, whorled in 2–5 cm long, **narrow clusters (spikes) at stem tip**, mid-summer.

FRUITS: Brownish, nut-like achenes, 4–5 mm long, oval, keeled (V-shaped); mid- to late summer.

WHERE FOUND: In deep lakes or large rivers; often in tea-coloured, nutrient-poor lakes with peaty shorelines; from Newfoundland to Lake Superior to central British Columbia, south to California and Georgia.

NOTES: Large-leaved pondweed is also known as 'musky weed' or 'bass weed.' • Large-leaved pondweed provides food for waterfowl and habitat for insects and fish. • The species name *amplifolius* means 'large leaves.'

FLOATING-LEAVED PONDWEED
POTAMOT FLOTTANT • *Potamogeton natans*

GENERAL: Perennial, aquatic herb; stems slightly flattened, 0.8–2 mm thick, usually unbranched; from rhizomes.

LEAVES: Alternate, parallel-veined, with leaf-like bracts (stipules); **submerged leaves bladeless**, stalkless, **10–40 cm long, 1–2 mm wide**, 3–5-veined; **floating leaves elliptic to oval, stalked, leathery, waxy on upper surface**, with many prominent veins.

FLOWERS: Small, with 4 sepal-like bracts, in **dense, narrow, 2–5 cm long clusters (spikes) at stem tip**; mid-summer.

FRUITS: Brownish, pitted, nut-like achenes, 3.5–5 mm long, shiny; mid- to late summer.

WHERE FOUND: Often forming extensive patches in shallow lakes and ponds, particularly on soft, organic, 'loonshit' bottoms; from Newfoundland to southern Alaska, south to California, Ohio and New Jersey.

NOTES: Floating-leaved pondweed is one of our most common pondweeds. It is usually easy to identify, but the leaf form can be quite variable. • **Oakes' pondweed** (*P. oakesianus*) is more delicate than floating-leaved pondweed. These 2 species are separated on the basis of their seeds, which are not pitted in Oakes' pondweed. Oakes' pondweed is found in pools, lakes and streams from southeastern Lake Superior to Newfoundland, south to Wisconsin and New Jersey. • Floating-leaved pondweed is a good food for waterfowl, and it provides habitat and food for fish.

VASEY'S PONDWEED
POTAMOT DE VASEY • *Potamogeton vaseyi*

GENERAL: Very **delicate**, perennial, **aquatic herb; stems thread-like, jointed**, often **rooting**, widely branching at base and with many short branches above, 30–45 cm long; of **2 forms**, emergent, fertile plants in shallow water and submerged, sterile plants in deep (2–2.5 m) water; sterile plants produce vegetative winter buds.

LEAVES: Of **2 types**, floating and submerged; **floating leaves leathery, opposite**, in 1–4 pairs, **egg-shaped**, widest above middle, 7–14 mm long, 3–6.5 mm wide, 5–9-nerved, slender-stalked, less common, found only on flowering plants; **submerged leaves thread-like**, 2.5–5 cm long, 0.1–0.5 mm wide; **stipules white, delicate**, 4–6 mm long, **free**.

FLOWERS: Tiny, lacking petals and sepals, in **dense, 3–5-flowered spikes** on thick, 6–12 mm long stalks; June–July.

FRUITS: Rounded achenes, about 2 mm long and almost as thick, 3-sided; mature spikes 4–6 mm long; August.

WHERE FOUND: In ponds and lakes at depths up to 5 m; from Ontario to New Brunswick, south to Minnesota, Illinois and Pennsylvania.

NOTES: Snailseed pondweed (*P. spirillus*, previously known as *P. dimorphus*) also has small, broad, leathery floating leaves and slender submerged leaves. However, its floating leaves are generally larger (up to 3 cm long and 12 mm wide) and fused (at least partly) to their stipules, and it has 2 distinct types of flower clusters—many-flowered, cylindrical spikes on erect stalks above the water, and few-flowered, rounded flower heads that are submerged. The fruits are twisted in a spiral like a tiny snail shell. Snailseed pondweed grows in shallow water of marshes, lakes, ponds and streams from South Dakota to Newfoundland, south to Iowa and Virginia.

VARIABLE-LEAVED PONDWEED
POTAMOT À FEUILLES DE GRAMINÉE
Potamogeton gramineus

GENERAL: Submerged perennial herb; stems long, flexible, semi-rounded, 0.5–1 mm thick, branched; anchored by roots and rhizomes.

LEAVES: Alternate, parallel-veined, with leaf-like bracts (stipules) all less than 3 cm long; **submerged leaves stalkless, 3–9 cm long, 0.5–1 cm wide,** 3–7-veined; **floating leaves stalked, broadly elliptic, 11–19-veined.**

FLOWERS: Small, with 4 sepal-like bracts, **whorled in narrow, 1.5–3 cm long clusters** (spikes) **at stem tips or from leaf axils**; mid-summer.

FRUITS: Green, nut-like achenes, 2–2.5 mm, sharply keeled (V-shaped); mid- to late summer.

WHERE FOUND: Often in shallower, more stagnant conditions than other pondweeds, such as fen pools and dense cattail stands; also in deeper lakes, ponds and river on a variety of substrates; from Newfoundland to Alaska, south to California, Ohio and New Jersey.

NOTES: Knotted pondweed (*P. nodosus*, previously known as *P. americanus*) is very similar to variable-leaved pondweed. Knotted pondweed, however, has stalked submerged leaves, and some of its stipules are over 3 cm long. It grows at depths of up to 2 m, in rivers and lakes from British Columbia to southern Labrador, south to Mexico, Alabama and Virginia. • The tubers of variable-leaved pondweed are an important waterfowl food, and the plants provide good food and shelter for fish. • The species name *gramineus* means 'grass-like.'

RIBBON-LEAVED PONDWEED
POTAMOT ÉMERGÉ • *Potamogeton epihydrus*

GENERAL: Submerged, perennial herb; stems long, flexible, somewhat flattened, 1–2 mm thick, up to 2 m long, few-branched; from rhizomes.

LEAVES: Alternate, with leaf-like bracts (stipules) and parallel veins; **submerged leaves ribbon-like, stalkless,** 5–20 cm long, 2–10 mm wide, **light-coloured with 2 mm wide bands on either side of midvein; floating leaves** 3–8 cm long, **oval to linear,** stalked.

FLOWERS: 2.5–4 mm wide, with 4 sepal-like bracts, in whorled **narrow, 1–3 cm long clusters (spikes) from leaf axils**; mid-summer.

FRUITS: Nut-like achenes, 2–3 mm long, keeled (V-shaped); mid- to late summer.

WHERE FOUND: In shallow (less than 2 m deep) lakes, ponds and streams; often found in fairly stagnant water; from Newfoundland to southern Alaska, south to California, Wisconsin and northern Georgia.

NOTES: Ribbon-leaved pondweed often grows with **variable-leaved pondweed** (*P. gramineus*, above). It provides food for waterfowl and muskrats. • The species name *epihydrus* means 'of the water surface.'

RICHARDSON'S PONDWEED
POTAMOT DE RICHARDSON • *Potamogeton richardsonii*

GENERAL: Robust, submerged, perennial herb; stems flexible, rounded, 12.5 mm thick, branching; from rhizomes.

LEAVES: Submerged, alternate, usually **less than 10 cm long and 2 cm wide**, lance- to egg-shaped, with **13–25 prominent parallel veins, wavy-edged, clasping, heart-shaped; bracts (stipules) at base 1–2 cm long, delicate, leaf-like**, soon shredding to white fibres and disappearing; young leaves bright green; **no floating leaves.**

FLOWERS: Small, with 4 sepal-like bracts, whorled in **narrow clusters (spikes) at stem tip, 1.5–3 cm long**; mid-summer.

FRUITS: Nut-like achenes, 2.5–3.5 mm long; mid- to late summer.

WHERE FOUND: In lakes, ponds and streams less than 4 m deep; from Newfoundland to Alaska, south to California, Ohio and North Carolina.

NOTES: This species was also known as *P. perfoliatus* ssp. *richardsonii*. • **White-stemmed pondweed** (*P. praelongus*) is very similar to Richardson's pondweed, but it has large (3–10 cm long), rigid, persistent stipules and large (4–5 mm long) achenes, and its rhizomes have noticeable, rust-coloured spots that are not found in Richardson's pondweed. Also, its leaves tend to be larger (at least 10 cm long and often more than 2 cm wide), with half-clasping bases and slightly hooded (rather than flat) tips. White-stemmed pondweed grows in deep water (up to 7 m) in lakes and ponds, and occasion-ally in rivers, from Alaska to Newfoundland, south to California, Indiana and New Jersey. • Richardson's pondweed is an important waterfowl food, and it also provides good habitat for insects and other invertebrates. • The genus name *Potamogeton* is an ancient Greek name, derived from the words *potamos*, 'river,' and *geiton*, 'neighbour,' in reference to the habitat of these plants.

FERN PONDWEED
POTAMOT DE ROBBINS • *Potamogeton robbinsii*

GENERAL: Perennial, submerged herb; stems up to 1 m long, branched above; from rhizomes lacking tubers.

LEAVES: Fan- or fern-like, alternate, linear, 3–10 cm long, 3–8 mm wide, parallel-veined with a **prominent midvein** and many smaller veins; **stipules fused to leaf blade; no floating leaves**.

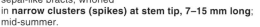

FLOWERS: Small, with 4 sepal-like bracts, whorled in **narrow clusters (spikes) at stem tip, 7–15 mm long**; mid-summer.

FRUITS: Nut-like achenes, 3.5–4.5 mm long, keeled (V-shaped); rarely produced, mid-summer.

WHERE FOUND: In deep or shallow lakes, ponds and rivers; forming extensive mats in deep lakes rich in organic detritus; often forms anchored mats; from Nova Scotia to Lake Superior and central British Columbia, south to California, Minnesota and Delaware.

NOTES: Fern pondweed is a poor food for waterfowl, but these plants provide cover for northern pike. The pondweeds are important food for moose, but little is known about individual species preference. • Fern pondweed can enhance wild rice production by adding organic material to the soil. In deep water, plants are often sterile.

FLAT-STEMMED PONDWEED
POTAMOT ZOSTÉRIFORME • *Potamogeton zosteriformis*

GENERAL: Perennial, submerged herb; stems **strongly flattened**, 1–3 mm wide, freely branched; from rhizomes; overwintering buds commonly produced.

LEAVES: Flat, alternate, 10–20 cm long, 2–5 mm wide, usually wider than stem, with 1–2 main veins and many parallel veins; **leaf-like bracts (stipules) present; no floating leaves.**

FLOWERS: Small, with 4 sepal-like bracts, whorled in cylindrical, 1.5–2.5 cm long, **narrow clusters (spikes) at stem tip**; mid-summer.

FRUITS: Nut-like achenes, 4–4.5 mm long, keeled (V-shaped); mid- to late summer.

WHERE FOUND: In shallow or deep quiet lakes, ponds and rivers with rich organic detritus; from Nova Scotia to Alaska, south to California, Indiana and Virginia.

NOTES: Flat-stemmed pondweed is occasionally eaten by waterfowl. • The species name *zosteriformis* means 'ribbon-like.'

175

SAGO PONDWEED
POTAMOT PECTINÉ • *Potamogeton pectinatus*

GENERAL: Submerged, perennial herb; stems rounded, 1 mm thick, 30–100 cm long, branching above; from rhizomes; **large overwintering buds sometimes produced in leaf axils.**

LEAVES: Submerged, alternate, with 3–5 parallel veins, stalkless, sharply pointed, 1–2 mm wide, with light-coloured bands on either side of midvein and leaf-like bracts (stipules) at base; **no floating leaves.**

FLOWERS: Small, with 4 sepal-like bracts, whorled in **dense, submerged, narrow, 1–4 cm long clusters (spikes) at stem tip**; mid-summer.

FRUITS: Nut-like achenes, 3–4.5 mm long, shiny, beaked; mid- to late summer.

WHERE FOUND: In shallow (less than 4 m deep) lakes, ponds and rivers; from Newfoundland to Alaska, south to California, Texas and Florida.

NOTES: Where common, the achenes, plants and tubers of this species are an important food for wildlife. Many waterfowl and marsh birds, including coots, black ducks, mallards, teals, wood ducks, Canada geese, scaups, redheads, common snipes and rails, fulfill up to half of their dietary needs with sago pondweed.

SLENDER PONDWEED
POTAMOT NAIN • *Potamogeton pusillus*

GENERAL: Slender, submerged, perennial herb; stems 0.1–0.7 mm thick, unbranched or branching, 20–150 cm long; from rhizomes; overwintering buds 1–3 cm long, sometimes produced in lower leaf axils in late summer, resembling much-shortened shoots with crowded, reduced leaves.

LEAVES: Submerged, alternate, stalkless, blunt, 3–30 mm long, 1–2 mm wide, with leaf-like bracts (stipules) and **2 glands at base; no floating leaves.**

FLOWERS: Small, with 4 sepal-like bracts, whorled in **narrow, 2–10 mm long, slender-stalked clusters (spikes) at stem tip**; mid-summer.

FRUITS: Nut-like achenes, shiny; mid- to late summer.

WHERE FOUND: In shallow (less than 2 m deep) lakes and ponds; from the maritime provinces to Alaska, south to California, Texas and Virginia.

NOTES: Slender pondweed is quite variable in leaf form, depending on water level and nutrient supply. This can cause identification problems, and a more detailed examination of the fruits is usually needed for positive identification. • **Leafy pondweed (*P. foliosus*)** is very similar to slender pondweed, but its leaves lack the 2 basal glands found in slender pondweed, and its small (usually 2-flowered) spikes are borne on short, club-shaped stalks. It grows in quiet, shallow water of ponds, lakes and rivers from the Northwest Territories to Labrador, south to Mexico and Florida. • Slender pondweed provides food for waterfowl and habitat and food for fish.

P. foliosus

SUBMERGED WATER STARWORT
CALLITRICHE HERMAPHRODITE
Callitriche hermaphroditica

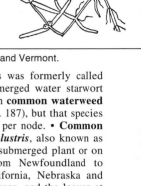

GENERAL: Delicate, submerged, annual or perennial herb; stems slender, limp, 5–30 cm long; typically solitary or in small patches.

LEAVES: Opposite, **linear**, 3–12 mm long, 0.5–1.3 mm wide, dark green, **all submerged** (no floating leaves).

FLOWERS: Minute, hidden in leaf axils, **no sepals or petals.**

FRUITS: 1–1.5 mm in diameter, consisting of 4 rounded lobes.

WHERE FOUND: Submerged in quiet waters of lakes and streams; from Newfoundland to Alaska, south to California, Minnesota and Vermont.

C. palustris

NOTES: This species was formerly called *C. autumnalis.* • Submerged water starwort could be confused with **common waterweed** (*Elodea canadensis*, p. 187), but that species typically has 3 leaves per node. • **Common water starwort** (*C. palustris*, also known as *C. verna*) occurs as a submerged plant or on wet, muddy soil, from Newfoundland to Alaska, south to California, Nebraska and Virginia. It is bright green, and the leaves at the stem tip are wider (broadly spoon-shaped) and often floating.

ALTERNATE-LEAVED WATER MILFOIL
MYRIOPHYLLE À FLEURS ALTERNES
Myriophyllum alterniflorum

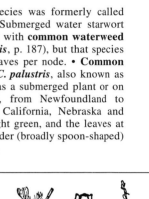

GENERAL: Perennial, **aquatic herb**; stems **very slender**, 15–20 cm long; from fibrous roots.

LEAVES: **Feather-like** (pinnately compound), with many, **thread-like, 5–12 mm long segments**, usually **in whorls** of 3–5.

FLOWERS: Inconspicuous, with 4 **short-lived, pale rose-coloured petals, solitary or in pairs in axils of small, short-lived, alternate leaves**; raised above water in 2–5 cm long spikes, with male flowers at tip and female flowers at base.

FRUITS: Hard nutlets, nearly 2 mm long, splitting into 4 1-seeded sections.

WHERE FOUND: In ponds and slow streams with soft organic bottoms; from Alaska to Newfoundland, south to Minnesota and Connecticut.

NOTES: Farwell's water milfoil (*M. farwellii*) differs from most of our water milfoils in that its flowers and fruits are borne in the axils of normal stem leaves, rather than in distinct spikes. It grows in quiet waters, usually on very soft, silty bottoms, from Ontario to Nova Scotia, south to Minnesota and New England. • **Slender water milfoil** (*M. tenellum*) is not immediately recognized as a water milfoil, because it lacks the feather-like leaves characteristic of this group. Its inconspicuous, undivided, thread-like leaves are often absent, and the unbranched, erect flowering stems are usually leafless, and they bear alternate flowers in either male or female spikes. Slender water milfoil grows in rich marshes, in shallow to deep water and on sandy to mucky shores, from Ontario to Newfoundland, south to Minnesota and New Jersey.

177

NORTHERN WATER MILFOIL
MYRIOPHYLLE DE SIBERIE • *Myriophyllum sibiricum*

GENERAL: Rooted or free-floating, submerged herb; stems white or light pink; **densely tufted, cylindrical, overwintering buds** (turions) produced **at branch tips** late in summer.

LEAVES: Feather-like, whorled, **1–5 cm long**, short-stalked, stiff, dark green, **pinnately divided, with 5–12 thread-like segments per side**.

FLOWERS: 1.5–3 mm long, **single in axils of upper leaves**, raised **above surface of water**; 2 tiny bracts at base of each flower usually shorted than flower, toothed or toothless (not feather-like); throughout summer.

FRUITS: Hard nutlets, 2 mm long, 4-lobed, eventually splitting into 4 1-seeded sections; mid- to late summer.

WHERE FOUND: In lakes, ponds, streams, quiet rivers, marshes and occasionally fen pools; from Newfoundland to Alaska, south to California, Texas, Ohio and Maryland.

NOTES: This species is also known as *M. exalbescens*. • Water milfoils are difficult to identify without flowers and fruits. • **Bracted water milfoil** (*M. verticillatum*) has feather-like bracts (divided into thread-like sections) that are usually more than twice as long as the flowers. Its stems are green or brown, most of its leaves lack stalks, and its overwintering buds are club-shaped. Bracted water milfoil is found from Newfoundland to Alaska, south to Utah, Indiana and Massachusetts. • **Eurasian water milfoil** (*M. spicatum*) is similar to northern water milfoil, but it has 12–20 leaf segments per side, and it does not produce overwintering buds. Eurasion water milfoil is an introduced weed of the eastern United States and southern Ontario, not yet common in northern Ontario. • Waterfowl and marsh birds eat the seeds and leaves of northern water milfoil in small quantities, and this plant is an important food for moose. The many invertebrates that live in these plants serve as food for fish and waterfowl. Northern water milfoil survives winter by producing overwintering buds. • The genus name *Myriophyllum* means 'with many leaves.'

M. sibiricum

M. verticillatum

M. verticillatum

M. verticillatum

TAPE GRASS, WILD CELERY
VALLISNÉRIE D'AMÉRIQUE • *Vallisneria americana*

GENERAL: Perennial, submerged herb; from fibrous roots; may form extensive beds from stolons.

LEAVES: Round-tipped, linear, up to 1 m long, 3–10 mm wide, **with 3 distinct bands** (1 many-nerved central band and 2 few-nerved or nerveless side bands); **in a basal rosette**.

FLOWERS: Small, solitary, **unisexual**; male flowers have 3 sepals and 1–2 petals, detach from plant, **rise to surface and float to female flowers**; female flowers have 3 sepals and 3 petals, **extend to surface** on **long, slender, coiled stalks**; mid-summer.

FRUITS: Cylindrical, 5–12 cm long, many-seeded; stalk coils to pull fruit underwater; late summer.

WHERE FOUND: In lakes, ponds and rivers; locally abundant; cosmopolitan; from New Brunswick to southern Manitoba, south to Texas and Florida.

NOTES: Tape grass leaves are distinguished from the submerged, tape-like leaves of **burreeds** (*Sparganium* **spp.**, pp. 163–64), **arrowheads** (*Sagittaria* **spp.**, pp. 160–61) and **bulrushes** (*Scirpus* **spp.**, pp. 142–46) by their wide midvein and narrow borders, which give them a 3-striped appearance. • Pollination occurs when male flowers are released to the water surface, where they float to reach female flowers. • All parts of tape grass plants are eaten by waterfowl, marsh birds and muskrats. This species is a favourite food of diving ducks, particularly canvasbacks and redheads, and it provides shade and shelter for fish and invertebrates. • Tape grass is not a grass, as the common name suggests. Rather, it is a member of the frog's-bit family (Hydrocharitaceae), more closely related to the water plantains (Alismataceae) and pondweeds (Potamogetonaceae) than to grasses (Poaceae).

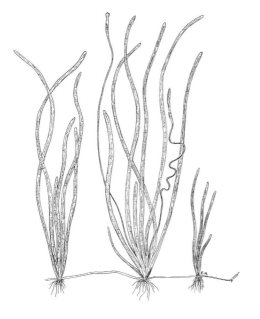

MERMAIDWEED
PROSERPINIE DES MARAIS • *Proserpinaca palustris*

GENERAL: Hairless, perennial, aquatic herb; stems simple, reclining at base, **flecked with tiny spicules**, 20–50 cm tall; rooting at lower nodes.

LEAVES: Alternate, **of 2 types; emergent leaves oblong to narrowly lance-shaped**, 2–5 cm long, 2–13 mm wide, sharp-toothed; **submerged leaves feather-like**, deeply cut into slender, pointed, minutely toothed segments.

FLOWERS: Small and inconspicuous, 3-parted, lacking petals, stalkless, 1–several in axils of emergent leaves; July.

FRUITS: Bony, 3–4-angled nutlets, rounded, about 2 mm in diameter, sharply 3-angled, concave on each side.

WHERE FOUND: Submerged in shallow waters or emergent along shores of marshes, ponds, lakes and streams, especially in lime-rich areas or where water levels fluctuate; from Ontario to Nova Scotia, south to Georgia and Mexico.

NOTES: Mermaidweed can be distinguished from **water milfoils** (*Myriophyllum* **spp.,** pp. 177–78) by its 3- rather than 4-parted flowers, its well-developed emergent leaves, and its alternate leaf arrangement. • Mermaidweed leaves vary remarkably with day length and degree of submergence. Submerged shoots always produce finely divided leaves, while aerial shoots produce divided leaves under long days but simple leaves under short days as summer progresses. High light intensity and high temperature can partly counteract the effects of submergence. The leaf forms can vary on a single stem if the plant is exposed to fluctuating water levels. • Mermaidweed seeds are eaten by ducks and muskrats.

COONTAIL
CORNIFLE NAGEANT • *Ceratophyllum demersum*

GENERAL: Submerged, aquatic herb, without roots; stems slender, branched.

LEAVES: 1–3 cm long, **split into 2 equal, thread-like segments, sharply toothed**, in whorls of 5–12 at each node, crowded near branch tips, giving the appearance of a racoon's tail.

FLOWERS: Very small, stalkless, either male or female; **solitary in axils of submerged leaves**; all summer.

FRUITS: Achenes, 4–5 mm long, **dark olive-green, elliptic, with 1 spine at tip and 2 spines at base**; mid- to late summer.

WHERE FOUND: Usually in lakes, ponds, marshes, streams and quiet rivers; can be found at depths of up to 7 m (deeper than other submerged plants); from Nova Scotia to Alaska, south to California, Texas and Florida.

NOTES: Coontail is easily distinguished from other submerged plants by the rough feel of its leaves. • **Water milfoils** (*Myriophyllum* **spp.,** pp. 177–78) have similar branches, but their leaves are pinnately divided. • Waterfowl eat coontail seeds and leaves, and muskrats occasionally eat the leaves. Coontail provides food and shelter for many invertebrates. • In quiet ponds, coontail can become very abundant, and it may form large beds.

FLAT-LEAVED BLADDERWORT
UTRICULAIRE INTERMÉDIAIRE • *Utricularia intermedia*

GENERAL: Aquatic, perennial herb; stems slender, floating, with **no roots**; **ends of branches often thick**, with dense leaves concealing overwintering buds (turions).

LEAVES: Alternate; up to 2 cm long, finely divided, flattened; **bladder-like traps are on leafless branches**.

FLOWERS: Showy, irregular, yellow, 2-lipped; lower lip 8–12 mm long (twice as long as upper lip); in clusters of 2–4, **elevated above surface on long stalk**; throughout summer.

FRUIT: 2-valved capsules, with many seeds; mid- to late summer.

WHERE FOUND: In marshes, fen pools and shallow ponds; typically in shallower water than common bladderwort; from Newfoundland to Alaska, south to California, Ohio and Delaware.

NOTES: Humped bladderwort (*U. gibba*) is very slender (its stems are less than 0.5 mm wide), and it creeps on the mud of shallow pools. Its leaves are very tiny, with only 2 small segments, and its yellow flowers are less than 1 cm long. It is found from Minnesota to Nova Scotia, south to California, Mexico and Florida. • The linear leaves of **inverted bladderwort** (*U. resupinata*) grow on or just below the soil surface on wet shores and in shallow water. This purple-flowered species grows in shallow water and on shores from Nova Scotia to northern Ontario, south to Illinois and Florida. • The small leaves of **horned bladderwort** (*U. cornuta*) grow in the soil, where they trap tiny soil organisms with their minute bladders. This yellow-flowered species grows in fens and on wet, sandy, muddy or peaty lakeshores from Newfoundland to northwestern Ontario, south to Texas, Minnesota, Delaware and Florida. • **Lesser bladderwort** (*U. minor*) is uncommon, and it forms mats on soil under shallow water from Newfoundland to Alaska, south to California, North Dakota and New Jersey. It has finely divided leaves up to 1 cm long, with 1–5 bladders on each, and the lower lip of its flowers is twice as long as the upper lip. • The genus name *Utricularia* means 'little bottle,' and it refers to the bladder-like animal traps.

U. cornuta

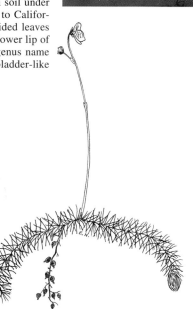

181

COMMON BLADDERWORT
UTRICULAIRE VULGAIRE • *Utricularia vulgaris*

GENERAL: Aquatic, perennial herb; free-floating, with **no roots**; **branch tips often thick**, with dense leaves concealing overwintering buds (turions).

LEAVES: Up to 5 cm long, finely divided, with **many bladder-like traps scattered among leaflets**; bladders often black when filled with prey.

FLOWERS: Showy, 1–2 cm long, **yellow, irregular, with 2 equal lips**; lower lip spurred; **elevated above water surface on long stalks**; throughout summer.

FRUITS: 2-valved capsules, with many seeds; mid- to late summer.

WHERE FOUND: Floating just below surface of water near lakeshores and in marshes, streams, ditches and shallow ponds; from Newfoundland to Alaska, south to California, Texas and Virginia.

NOTES: Hidden-fruited bladderwort (*U. geminiscapa*) and purple bladderwort (*U. purpurea*) are also free-floating, and they have bladders scattered among their leaf segments. In addition to its showy, yellow blooms (seldom produced), hidden-fruited bladderwort produces small, self-fertilizing flowers that lack petals, and that never open. It is rare in Ontario, but it grows in ponds or slow-moving streams from Wisconsin to Newfoundland, south to Virginia. Purple bladderwort has showy, purple flowers, and its leaves are finely divided into whorls of hair-like segments, each tipped with a tiny bladder. It grows in ponds or slow-moving streams from Wisconsin to Newfoundland, south to Louisiana and Florida. Both of these species are rare and local in southern and central Ontario.

• The bladders on the leaves have hair-triggered doors that open when tiny aquatic animals brush against them. The water then rushes into the empty bladder, carrying the animal with it for the plant to digest and absorb. Some bladderworts have simple or only slightly forked leaves. • Common bladderwort is eaten sparingly by moose, waterfowl and muskrats.

U. purpurea   *U. geminiscapa*

CURLY WHITE WATER CROWFOOT
RENONCULE À LONG BEC • *Ranunculus longirostris*

GENERAL: Perennial, submerged herb; stems floating, curly, flexible, unbranched or few-branched, rooting from lower nodes.

LEAVES: Stiff (do not collapse when taken out of water), short-stalked, 1–2 cm long, **finely divided** into **tubular, thread-like segments.**

FLOWERS: Slightly raised above surface of water, 1–1.5 cm wide, with **5 purplish-green sepals and 5 white petals**, plus yellow stamens at centre; early to mid-summer.

FRUITS: Beaked achenes, in a hemispherical head; mid- to late summer.

WHERE FOUND: In lakes, ponds, streams and rivers; from Newfoundland to Alaska, south to California, Texas and Delaware.

NOTES: Yellow water crowfoot (*R. flabellaris*) and small yellow water crowfoot (*R. gmelinii*), which both have yellow flowers and flattened, thread-like leaf segments, grow in ponds, streams and wet meadows and along shores across much of North America. • Non-flowering crowfoot plants can be distinguished from water marigold (*Megalodonta beckii*, p. 184) by their short leaf stalks. Water marigold leaves are stalkless. • Crowfoots contain a poisonous substance, and they should not be eaten. • All species of *Ranunculus* contain anemonin and protoanemonin which are antiseptic against bacteria. These plants have been used as a treatment for rheumatism and as an ancient remedy for cancer. • Curly white water crowfoot seeds are occasionally eaten by moose, waterfowl and marsh birds. The leaves provide habitat for small freshwater animals and insects that in turn provide food for fish. • The genus name *Ranunculus* means 'little frog,' and it refers to the amphibious habitat of many species of this genus.

R. gmelinii

CREEPING SPEARWORT
RENONCULE RAMPANTE • *Ranunculus reptans*

GENERAL: Low, trailing, perennial herb; **stems creeping, rooting** at joints.

LEAVES: Linear to thread-like, 6–25 mm long.

FLOWERS: Bright **yellow**, about 8–10 mm wide with 4–7 showy petals; **solitary** on 2.5–7.5 cm long stalks.

FRUITS: Flattened achenes, with tiny, sharp beaks.

WHERE FOUND: On damp sandy or gravelly lakeshores and streambanks, less often on muddy, silty, mucky or rocky shores; from Alaska to Newfoundland, south to California, Colorado, Minnesota and Massachusetts.

NOTES: This species is also known as *R. flammula*. • When not in flower, creeping spearwort could be confused with a small rush. Sterile, aquatic plants are easily recognized by their arching, green stolons, with thread-like leaves that are squared at the tip. • Many buttercups are eaten by moose and aquatic birds, but the juice of most species can cause **skin irritation and blistering** in humans. These toxic chemicals are rendered harmless by drying or boiling. • Creeping spearwort was traditionally used to treat blisters and ulcers. • The genus name *Ranunculus* is the dimunitive of *rana*, 'a frog,' and it was given to these plants because many species grow in damp places. The species name *reptans* means 'creeping.'

WATER MARIGOLD
BIDENT DE BECK • *Megalodonta beckii*

GENERAL: Submerged, perennial herb; stems hairless, thicker below surface of water.

LEAVES: Opposite, stalkless; **emersed leaves lance-shaped, toothed,** 2–4 cm long; **submerged leaves 2–4 cm long, finely dissected into thread-like segments**.

FLOWERHEADS: Golden-yellow, sunflower-like, 3–5 cm wide, with 6–10 ray florets around a central disc, lifted above surface of water; late summer.

FRUITS: Rounded achenes, with 3–6 long, slender bristles that are barbed near tips.

WHERE FOUND: In quiet ponds and streams; from Nova Scotia to southern British Columbia, south to Missouri, Pennsylvania and New Jersey.

NOTES: This species was previously known as *Bidens beckii*. • The submerged leaves may be confused with those of **coontail** (*Ceratophyllum demersum*, p. 180), **bladderworts** (*Utricularia* **spp.**, pp. 181–82), **water milfoils** (*Myriophyllum* **spp.**, pp. 177–78), or **water crowfoots** (*Ranunculus* **spp.**, p. 183), but the leaves of water marigold differ in that they are toothless, opposite and lacking bladders and stalks. Water marigold stems tend to be thicker than those of other submerged plants. The pictorial key to aquatic species and the plant descriptions will help you to distinguish them. • Ducks occasionally eat water marigold achenes.

LAKE CRESS
ARMORACIA LACUSTRE • *Armoracia lacustris*

GENERAL: Perennial, **aquatic herb**, with fibrous roots; flowering stems 30–60 cm long, **submerged or prostrate**.

LEAVES: Submerged leaves, 5–15 cm long, 1–3 times deeply cut into **thread-like segments**; stem leaves above water stalkless to short-stalked, **oblong, finely to coarsely toothed**.

FLOWERS: White, showy, 4-parted, on slender, widely spreading, 6–8 mm long stalks; in elongated clusters (racemes) at stem tips.

FRUITS: Ovoid pods (silicles), to 8 mm long, with a 4 mm long style; seeds few, in 2 rows.

WHERE FOUND: In quiet waters of lakes and slow streams, especially if cold and spring-fed, and also on muddy shores; from Ontario to southwestern Quebec, south to Louisiana and Florida.

NOTES: This species was previously known as *Rorippa aquatica, A. aquatica, Neobeckia aquatica* and *Radicula aquatica*. • Lake cress can by distinguished from **coontail** (*Ceratophyllum demersum*, p. 180), **water marigold** (*Megalodonta beckii*, p. 184) and most **water milfoils** (*Myriophyllum* **spp.**, pp. 177–78) by its alternate leaves. • **Mermaidweed** (*Proserpinaca palustris*, p. 180) and some water milfoils have alternate leaves, but the side segments of their leaves are not as finely divided as those of lake cress. • The presence of a central leaf axis distinguishes lake cress from **curly white water crowfoot** (*Ranunculus longirostris*, p. 183) and the **bladderworts** (*Utricularia* **spp.**, pp. 181–82). • Another common wetland mustard is **marsh yellow cress** (*Rorippa palustris*, also known as *R. islandica*). It has many small (usually less than 6 mm wide), bright yellow, 4-parted flowers and long-stalked, short-cylindrical to egg-shaped pods in elongated clusters (racemes). The leaves of marsh yellow cress are deeply cut into toothed lobes, with the largest lobe at the leaf tip. Its submerged leaves are not as finely divided as those of lake cress, however, and it has a taproot rather than fibrous roots. Marsh yellow cress grows in rich deciduous swamps, thickets, shallow water and ditches and on wet shores from Alaska to Newfoundland, south to California, Texas and Florida. • The finely divided, submerged leaves of lake cress fall off readily when mature; simply lifting the plant from the water may cause them to drop. If fallen leaves and stem fragments settle in a suitable site, they can grow into new plants. Lake cress often grows in profusion, crowding out other aquatic plants. Lake cress belongs to the same genus as horse radish (*A. rusticana*). • The species name *lacustris* means 'of the lake.'

R. palustris

R. palustris

SLENDER NAIAD, WATER NYMPH
NAÏADE FLEXIBLE • *Najas flexilis*

GENERAL: Submerged, annual herb, tall (up to 50 cm) and slender, with forked branches or short (10–20 cm tall) and bushy; from fibrous roots.

LEAVES: Opposite, 0.2–0.6 mm wide, ascending, with very small teeth; **broad at base and tapering to slender point**.

FLOWERS: Very small, solitary, stalkless, in leaf axils, often hidden by broad, sheathing leaf bases; male and female flowers separate but on same plant; mid-summer.

FRUITS: Achenes, 2.5–3.7 mm long, with a papery sheath enclosing a glossy, yellow to brown seed; mid- to late summer.

WHERE FOUND: In lakes, ponds, rivers and streams; from Newfoundland to the Northwest Territories, south to California, South Dakota and North Carolina.

NOTES: Thread-like naiad (*N. gracillima*) has whorls of 3–5 thread-like leaves with distinctive, broad, lobes (auricles) at their base. It has pitted, dull, cylindrical seeds. Thread-like naiad grows in shallow, mucky-bottomed lakes, from Minnesota to eastern Ontario and Nova Scotia, south to Missouri and Virginia. It is intolerant of pollution, and it is rare and local in Ontario. • **Common waterweed** (*Elodea canadensis*, p. 187) is similar to slender naiad, but its leaves are wider (2.5–3 mm wide) without broad bases. • Waterfowl, such as black ducks, buffleheads, pintails, ring-necked ducks, scaups, teals, geese and coots, eat the plants and seeds. Slender naiad also provides food and shelter for fish. • The genus name *Najas* means 'water-nymph.'

PIPEWORT
ÉRIOCAULON AQUATIQUE • *Eriocaulon aquaticum*

GENERAL: Perennial, aquatic herb, with a single, long-stalked, button-like flowerhead; from white, fibrous, fleshy, **distinctively segmented**, unbranched roots.

LEAVES: 2–5 mm wide, 2–10 cm long, translucent, **with parallel veins and 3–9 conspicuous cross-veins**, pale and spongy at base; in basal rosettes.

FLOWERS: Minute, with **hairy tepals** and basal bracts, male or female on same or separate plants; in single, tight flowerheads 4–6 mm in diameter, on 3–20 cm long (longer in deep water), ridged stalks, usually raised above water; mid- to late summer.

FRUITS: Small, thin-walled capsules, with oval seeds; late summer.

WHERE FOUND: In acidic lakes with sandy or peaty shorelines and on floating fen mats; often forms dense turfs in shallow water; from Newfoundland to northwestern Ontario, south to Indiana, Ohio and Delaware.

NOTES: This species was previously known as *E. septangulare*. • Pipewort is one of our most common rosette-forming aquatics. It is often confused with **quillworts** (*Isoetes* **spp.**, p. 157), but pipewort can be distinguished by its distinctively segmented roots.

COMMON WATERWEED
ELODÉE DU CANADA • *Elodea canadensis*

GENERAL: Submerged, perennial, aquatic herb; **stems rounded, branched**; overwintering buds (short compact branches) produced in late summer.

LEAVES: Oval (on female plants) or lance-shaped (on male plants), 6–17 mm long, bright green when young, opposite near base of stem, **in whorls of 3 in middle and upper part of stem**; upper leaves closely overlapping; young leaves bright green.

FLOWERS: Up to 9 mm in diameter**, white, unisexual**, with male or female plants (**dioecious**), in upper leaf axils; female flowers raised to surface of water by **3–20 cm long, thread-like stalks**; male flowers release buoyant pollen that drifts to floating female flowers; throughout summer.

FRUITS: Oval capsules, beaked, 6 mm long; 6 seeds, narrow, cylindrical; mid- to late summer.

WHERE FOUND: In lakes, ponds, marshes and rivers; sometimes common in lakes with low nutrient availability and in deep, cold bays in Lake Superior; from Nova Scotia to southern British Columbia, south to California, Iowa and North Carolina.

NOTES: This species was formerly known as ***Anacharis canadensis***. • Common waterweed is easily distinguished from other submerged plants by its dense, short, oval, bright green, opposite/whorled leaves. It is one of the few submerged vascular plants found at depths greater than 10 m. • The genus name *Elodea* means 'of marshes.'

COMMON MARE'S TAIL
HIPPURIDE VULGAIRE • *Hippuris vulgaris*

GENERAL: Submerged or emergent, perennial herb; **stems thick, soft, unbranched, erect, 10–50 cm tall**; from spongy rhizomes.

LEAVES: Unbranched, **in whorls of 6–12**; emergent leaves 1–2 cm long, thick, firm, spiky; **submerged leaves 1–3 cm long, thin and weak**, drooping when removed from water.

FLOWERS: Very small, **lacking sepals and petals**, in axils of upper leaves; **rare**, summer.

FRUITS: Nutlet, ellipsoidal, 1.7–2.5 mm long; mid- to late summer.

WHERE FOUND: In lakes, marshes, streams and rivers; sometimes on peat in fens and often in shallow marshes near bogs or shallow meandering, spring-fed streams; from Labrador to Alaska, south to California, Nebraska and New England.

NOTES: A saltwater species, **four-leaved mare's tail** (*H. tetraphylla*), grows on wet, saline sites, usually on coastlines, from Newfoundland to Alaska, south to central British Columbia and James Bay. • Water-fowl and marsh birds occasionally feed on the leaves and stems of common mare's tail. • *Hippuris* means 'horse's tail.'

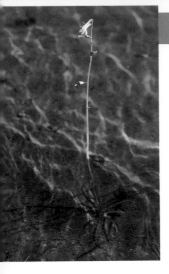

WATER LOBELIA
LOBÉLIE DE DORTMANN • *Lobelia dortmanna*

GENERAL: Emergent perennial herb; stems up to 1 m tall, slender, hollow, rising above surface of water; from fibrous roots.

LEAVES: Submerged, linear, tubular, fleshy, in a 2–9 cm wide basal rosette; stem leaves few, minute, very thin.

FLOWERS: Showy, violet or white, 1–2 cm long, irregular, sparsely arranged in elongated cluster at stem tip (raceme); mid- to late summer.

FRUITS: Capsules, splitting open from tip to release seeds; late summer to early autumn.

WHERE FOUND: On lakeshores with sandy, muddy or silty bottoms, in up to 50 cm of water; often found in nutrient-poor, acidic lakes; from Newfoundland to southern British Columbia, south to Pennsylvania and New Jersey.

NOTES: Water lobelia often grows with **quillwort** (*Isoetes* spp., p. 157) and **pipewort** (*Eriocaulon aquaticum*, p. 186). • **Kalm's lobelia** (*L. kalmii*) has blue to white flowers and flat, rather than tubular, leaves. It is found in calcium-rich habitats, including edges of lakes, fens and wet meadows from Newfoundland to northern British Columbia, south to Washington, South Dakota and New Jersey. • **Cardinal flower** (*L. cardinalis*) is a terrestrial rather than aquatic species with large (60–120 cm tall), leafy stems and large (2.5–4 cm long), bright scarlet flowers. It grows in damp meadows and along streambanks from Minnesota to New Brunswick, south to eastern Texas and Florida. • *Lobelia* is named after M. de l'Obel, a Renaissance pioneer of botany and the herbalist for James I in the late 1500s.

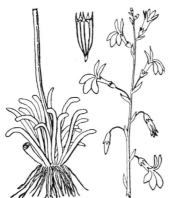

L. cardinalis   *L. kalmii*

STONEWORT, MUSKGRASS
LUSTRE D'EAU • *Chara* spp.

GENERAL: Algae that resemble vascular plants, with cylindrical, whorled branches, often encrusted with lime, usually with a **musky smell** similar to rotting garlic.

BRANCHES: Whorled at nodes; **rough**, 12 cm long, usually with fine lines.

REPRODUCTIVE STRUCTURES: Visible only with a microscope; female structures (oogonia) reddish, **capped by 5 cells**; male structures (antheridia) produce sperm with tails that can swim to oogonia.

WHERE FOUND: On bottoms of quiet lakes, ponds, stream pools and slow rivers; most abundant in calcium-rich waters, such as marl lakes; across North America.

NOTES: Although they resemble vascular plants, stoneworts are filamentous algae more closely related to the green algae ('frog spit') that forms a scum on the surface of ponds. Each joint of the stem consists of a single cell. • *Nitella* spp. are similar algae, but they have smooth branches, their oogonia are capped by 10 cells, and they lack the musky smell. They also grow in quiet waters, across North America. • Stoneworts are a primary food for waterfowl, such as coots, black ducks, teals, mallards and goldeneyes, especially when the oogonia are produced. Ducks also eat the small freshwater animals that live on these plants. Moose occasionally eat stonewort.

FRESHWATER SPONGES

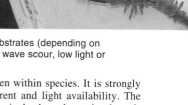

GENERAL: Often mistaken for plants, but actually colonies of simple, submerged, aquatic, filter-feeding **animals; highly variable, from finger-like or branching** structures (up to 30 cm long) to flat or convoluted encrustations on substrate; often **bright green** due to presence of symbiotic algae, but also grey, white or brown.

WHERE FOUND: Attached to rock, wood, or plants in lakes, ponds and slow-moving rivers, on either hard or soft substrates (depending on species); can be limited by low pH, low mineral concentration, wave scour, low light or oxygen levels and low temperatures; across North America.

NOTES: The growth form of freshwater sponges varies, even within species. It is strongly influenced by environmental conditions, such as water current and light availability. The body of a freshwater sponge is made of a skeleton of silica spicules bound together by collagen. Species identification requires microscopic examination of the spicules. • Sponges reproduce sexually when the males release sperm into the water, to be taken up by the females. They also reproduce asexually by fragmentation, and by producing specialized packets of cold-resistant cells called 'gemmules.' Gemmules are usually produced internally in the autumn, and as the parent sponge disintegrates over the winter, they are released and carried by water currents to form new sponges the following spring. • Approximately 30 species of freshwater sponges, in 8 genera, are found in eastern North America. *Ephydatia muelleri* and *Spongilla lacustris* are the 2 most common freshwater sponges in northern Ontario, though other species are occasionally found.

FERNS AND ALLIES

This group includes the ferns, horsetails, clubmosses and quillworts. Pteridophytes differ from herbs in that they reproduce by spores (rather than seeds), and they differ from bryophytes in that their sporophytes are separate from their gametophytes for part of their life cycle. This places pteridophytes between the higher plants (trees, shrubs and herbs) and the lower plants (mosses and liverworts). Ferns and their allies require moisture for fertilization to occur, thus restricting them to moist habitats.

The key to identifying plants in this group is to have a good understanding of the leaf parts and the terminology for compound leaves (see inside back cover). The labelled drawings below should help.

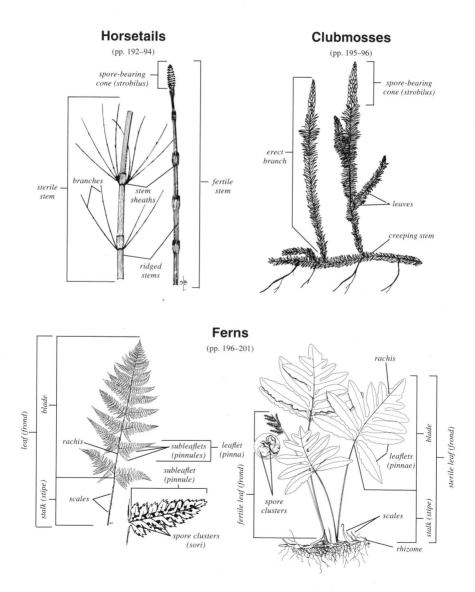

Horsetails
(pp. 192–94)

Clubmosses
(pp. 195–96)

Ferns
(pp. 196–201)

191

WOOD HORSETAIL
PRÊLE DES BOIS • *Equisetum sylvaticum*

GENERAL: Perennial horsetail, up to 50 cm tall, with regular whorls of delicate, arching, **branched branches**; stems jointed, hollow, with 10–18 rough, vertical ridges; mature fertile and sterile stems similar; **fertile stems unbranched and pinkish to whitish at first**, turning green and developing branches after spores are shed; **sterile stems green, with many soft branches**; from dull black rhizomes.

LEAVES: Reduced to small, whorled scales at stem and branch joints (nodes), fused to form sheaths; stem sheaths loose and flaring, reddish brown near tip, with many teeth joined to form 3–4 broad, papery lobes.

SPORE CLUSTERS: In long-stalked, blunt-tipped cones (strobili) at tips of fertile stems; spores shed in early spring to mid-summer, cones shed soon afterward.

WHERE FOUND: In swamps and moist forests and on streambanks; from Alaska to northern Labrador and Newfoundland, south to Idaho, South Dakota and Virginia.

NOTES: Wood horsetail is our only species of horsetail with branched (compound) branches. • Horsetails are high in silicates, and they have been used as bottle brushes and pot scrubbers. Native people made a tea from horsetails to treat kidney and bladder trouble. Horsetails contain a **toxin** that destroys vitamin B_1, and they are **poisonous to horses, sheep and cattle** when eaten in large quantities. • Horsetails dominated the landscape in the carboniferous period (about 300 million years ago), and they make up a high proportion of coal deposits. Fossilized horsetails, called 'calamites,' have been found with stems 20 cm in diameter and 15 m tall. • The genus name *Equisetum* means 'horse hair,' and it refers to the slender branches of many species.

MEADOW HORSETAIL
PRÊLE DES PRÉS • *Equisetum pratense*

GENERAL: Perennial horsetail, 20–50 cm tall; stems jointed with 12 or more, rough, vertical ridges and **many regular whorls of soft, simple branches**; mature fertile and sterile stems similar; **fertile stems unbranched and whitish at first, turning green and developing branches** after spores are shed; sterile stems, green and branched; lowest branch segment (first internode) shorter than adjacent stem sheath; from dull black rhizomes.

LEAVES: Reduced to small, whorled scales at stem and branch joints (nodes), fused to form sheaths; stem sheaths with **short, pointed teeth** with thin, white edges; branch sheaths with 3 tiny teeth.

SPORE CLUSTERS: In 2–3 cm long cones (strobili) at tip of green, fertile stems that develop whorls of branches; shed spores in early spring to mid-summer.

WHERE FOUND: In moist forests, swamps, ditches and wet meadows and on shaded streambanks; from Alaska to Newfoundland, south to Montana, Iowa and New Jersey.

NOTES: Field horsetail (*E. arvense*) is very similar to meadow horsetail, but its lowest branch segments are equal to or longer than the adjacent stem sheaths. The fertile stems of field horsetail appear early in the spring, they are small (10–20 cm tall) and unbranched, and they lack chlorophyll. The typical green horsetail plant, with its many whorls of branches, is then produced later in the spring. Field horsetail usually grows on drier sites than meadow horsetail, and it is found from Alaska to northernmost Ellesmere Island, south through most of Canada and the U.S. • **Marsh horsetail** (*E. palustre*) differs from these 2 horsetails in that it produces spores in the summer (rather than in spring), its branch sheaths have 6–8 (rather than 3) teeth, and the central cavity of its stem is $1/6$ the diameter of the stem (vs. $1/3$ the diameter in meadow horsetail and $1/4$ the diameter in field horsetail). It grows in wet meadows and woods, in marshes and on shores from Alaska to Newfoundland, south to Idaho, Nebraska and New Jersey.

E. arvense, fertile stems

E. arvense, sterile stems

WATER HORSETAIL
PRÊLE FLUVIATILE • *Equisetum fluviatile*

GENERAL: Aquatic, perennial horsetail; **stems often over 1 m tall**, consisting of 5 cm long segments, **soft (collapsing easily when squeezed), hollow**, with a central cavity $4/5$ the diameter of the stem, and **approximately 20 vertical ridges**; branches none to many, irregularly to regularly whorled; from smooth branching rhizomes, sometimes forming dense patches.

LEAVES: Reduced to small, whorled scales at stem and branch joints (nodes), fused to form sheaths; **stem sheaths tightly appressed, dark brown, with 15–20 narrow teeth.**

SPORE CLUSTERS: In **yellow or brown cones** (strobili) **at tip of stem, 2–5 cm long, rounded at tip**; throughout summer.

WHERE FOUND: In lakes, fens, marshes, streams, swamps, ditches and wet meadows; most abundant in shallow water by sand bars and silty shores; from Alaska to Labrador, south to Washington and Virginia.

NOTES: Two slender *Equisetum* species never produce whorls of side branches. **Dwarf scouring-rush** (*E. scirpoides*) is very small, with 7–15 cm long, spreading, thread-like stems and 3-toothed sheaths. It grows in swamps and especially moist coniferous forests, from Alaska to Newfoundland, south to Washington, South Dakota and New York. **Variegated scouring-rush** (*E. variegatum*) is 15–45 cm tall, and its sheaths have 5–10 teeth. The cone has a sharp tip. It grows on sandy lakeshores and riverbanks, and in ditches and wet meadows, from Alaska to northernmost Ellesmere Island and Labrador, south to California, Illinois and New Jersey. • **Marsh horsetail** (*E. palustre*, p. 193) can be similar to water horsetail, but it has only 6–8 teeth on its stem sheaths. • Birds and mammals, such as geese, grouse, swans, muskrats and black bears, eat water horsetail rhizomes and stems. Moose occasionally feed heavily on these stems. • The species name *fluviatile* means 'of rivers.'

E. scirpoides

NORTHERN BOG CLUBMOSS
LYCOPODIELLE INCONDÉ • *Lycopodiella inundata*

GENERAL: Creeping, evergreen, perennial clubmoss; stems trailing, 5–20 cm long, with erect, 3–10 cm tall fertile branches; from adventitious roots.

LEAVES: Linear, pointed, about 6 mm long, crowded in 8–10 vertical rows, twisted towards upper side of stem.

SPORE CLUSTERS: In axils of crowded, broad-based leaves (sporophylls) at tips of unbranched, fertile shoots; late summer.

WHERE FOUND: In fens and on sandy lakeshores; from Alaska to Idaho and California in the west, and from Ontario to Newfoundland, south to Texas and Florida in the east.

NOTES: This species is also known as *Lycopodium inundatum*. • **Stiff clubmoss** (*Lycopodium annotinum*) grows in swamps and moist conifer forests from Alaska to Baffin Island and Labrador, south to Oregon, Colorado, Minnesota and Virginia. It differs from northern bog clubmoss in that its spore-bearing leaves (sporophylls) are reduced to yellow-brown scales, and they form distinct cones (strobili) at the tips of erect or ascending, branching shoots. There are about 12 other species of *Lycopodium* in Ontario. • Clubmosses are not 'mosses.' They are vascular plants with large, spore-producing structures. Clubmosses are larger, coarser and stiffer than true mosses. • The yellow spores of clubmosses were once used as flash powder for cameras and in explosives. • The species name *inundatum* means 'apt to be flooded.'

Lycopodiella inundata

Lycopodium annotinum

Lycopodium annotinum

195

BOG SPIKEMOSS
SÉLAGINELLE FAUSSE-SÉLAGINE • *Selaginella selaginoides*

GENERAL: Inconspicuous, **yellowish-green, moss-like**, perennial spikemoss; sterile stems slender, **creeping**, 1–2 cm long; fertile stems thicker, erect to ascending, 3–8 cm tall.

LEAVES: Lance-shaped, pointed, 2–4 mm long, with small, scattered **bristles along edges**.

SPORE CLUSTERS: In axils of lance-shaped, fringed bracts, densely clustered at branch tips to form solitary, **narrowly oblong cones less than 2 cm long**.

WHERE FOUND: In fens and damp woods and on moist shores, banks and wet talus slopes, on neutral to slightly alkaline soil; from Alaska to Newfoundland, south to Nevada, Wyoming, Minnesota and Maine.

NOTES: Bog spikemoss often grows among mosses. • **Creeping spikemoss** (*S. apoda*) has leaves of 2 shapes, arranged in 4 ranks; leaves in the 2 side rows are bluntish, oblong to oval and spreading; dorsal and ventral leaves are pointed, smaller and appressed. The leaves and sporophylls of creeping spikemoss lack cilia, and the fertile branches are stalkless. This species grows in wet woods, swamps, meadows and damp lawns and on streambanks and shores, on neutral to subacid soil. • These primitive vascular plants are most closely related to the **clubmosses** (*Lycopodium* spp., p. 195) and **quillworts** (*Isoetes* spp., p. 157). • Resurrection plant (*S. lepidophylla*), which is found in dry areas in the southwestern U.S. and Mexico, is often sold as a novelty, since it curls up, turns brown and appears dead when dry, but it uncurls and regreens when moistened.

RATTLESNAKE FERN
BOTRYCHE DE VIRGINIE • *Botrychium virginianum*

GENERAL: Erect, perennial grape fern, 20–75 cm tall.

LEAVES: Solitary, **triangular**, 7–20 cm wide, 10–30 cm long, stalkless, 3-parted; 3 main parts 2–3 times pinnately divided, lacy and delicately toothed.

SPORE CLUSTERS: Numerous, yellow, rounded, in dense, **grape-like clusters** 6–15 cm long at tip of solitary, fertile stalks (7–20 cm long) from junction of 3 green (sterile) leaflets (pinnae); throughout summer.

WHERE FOUND: In rich swamps, moist forests and shrubby thickets and on shaded streambanks; from Newfoundland to Alaska, south to California, Indiana and Maryland.

NOTES: The distinctive spore clusters make rattlesnake fern easy to identify. • Rattlesnake fern is sometimes confused with **oak fern** (*Gymnocarpium dryopteris*, p. 197), but that fern does not have a separate fertile leaf, and the smallest lobes of its leaves are rounded, not delicately toothed. • **Bracken fern** (*Pteridium aquilinum*) also has leaves with 3 main divisions, but its leaves are much larger (20–90 cm wide, on stalks 20–90 cm tall) and coarser, and its spores are borne in a continuous band along the edges of the leaves. Bracken fern grows in drier sites, such as open woods and burned areas, from southern Alaska to Newfoundland, south to California, Texas and North Carolina. • The genus name *Botrychium* means 'bunch of grapes,' and it refers to the spore clusters on the fertile branches.

OAK FERN
DRYOPTÈRE DISJOINTE • *Gymnocarpium dryopteris*

GENERAL: Perennial fern, 10–30 cm tall; from **slender, black rhizomes**.

LEAVES: Triangular, almost hairless, 18 cm long, 25 cm wide, divided into 3 main leaflets (pinnae), with upper pinna largest; main leaflets **twice pinnately divided, with smallest leaflets (pinnules) bluntly lobed**; stalks thin, dark, stiff with a few scales at base.

SPORE CLUSTERS: Small, rounded, lacking a membranous covering (indusium); **on underside of leaflets near edges**; mid-summer.

WHERE FOUND: In swamps and moist forests and on moist, shaded cliffs; from Alaska to Newfoundland, south to Arizona, Iowa and Virginia.

NOTES: Other ferns with leaves divided into 3 main parts include **rattlesnake fern** (*Botrychium virginianum*, p. 196), which has a specialized fertile branch and delicately toothed leaves, and **bracken fern** (*Pteridium aquilinum*), which has larger (up to 1 m tall), coarser leaves. • The delicate leaves of **bulblet fern** (*Cystopteris bulbifera*) are 30–60 cm long, twice pinnately compound and lance-shaped in outline, with long-tapering tips. They often bear small bulblets on their lower side, which fall to the ground and grow into new plants. Bulblet fern is occasionally found in wetlands, but it is more common in moist woods and on calcareous rocks from Ontario to Newfoundland, south to Arizona, Arkansas and Georgia. • The genus name *Gymnocarpium* means 'naked fruit,' and it refers to the spore dots (sori), which lack a membranous covering (indusium).

G. dryopteris

Cystopteris bulbifera

Cystopteris bulbifera

197

SENSITIVE FERN
ONOCLÉE SENSILE • *Onoclea sensibilis*

GENERAL: Erect, perennial fern; from rhizomes.

LEAVES: Of 2 distinct types (sterile and fertile), in clusters with sterile leaves surrounding fertile leaves; sterile leaves 50–70 cm tall, **once pinnately divided into opposite lobes, with wavy edges** and a winged central axis; **fertile leaves described with spore clusters.**

SPORE CLUSTERS: Enclosed in **tightly rolled, berry-like bodies** (leaflets) of fertile leaves; fertile leaves 30–75 cm tall, dark brown; late summer, spores shed late in autumn, **persisting through winter.**

WHERE FOUND: In black ash and cedar swamps, on shores, around edges of beaver ponds and in wet meadows; from southern Manitoba to Newfoundland, south to Texas and Florida.

NOTES: The green, sterile leaves of sensitive fern are **poisonous to horses**, and they can present a problem if they are incorporated into bales of hay. • Sensitive fern is sometimes cultivated as an ornamental plant, and it can spread rapidly, becoming a troublesome weed. • The rattle-like fertile leaves have a unique shape and texture, and they are often collected by children. • *Onoclea* means 'closed cup,' and it refers to the spore clusters (sori), which are enclosed in the curled leaflets of fertile leaves. The species name *sensibilis* means 'capable of irritability or sensitivity,' and it refers to the sterile leaves, which are very sensitive to early frost.

LADY FERN
ATHYRIE FOUGÈRE-FEMELLE • *Athyrium filix-femina*

GENERAL: Delicate, showy, perennial fern, up to 1 m tall, **tufted**; from **scaly**, creeping **rhizomes**.

LEAVES: 40–100 cm long, 10–35 cm wide, **broadly lance-shaped, twice (sometimes 3 times) pinnately divided; stalks brittle, brown-scaly, shorter than blade**; smallest leaflets (pinnules) lobed.

SPORE CLUSTERS: Long, narrow, curved to horseshoe-shaped (a hand lens is helpful), on underside of leaf; summer.

WHERE FOUND: In black ash or eastern white cedar swamps, moist forests and wet fields, along streams and in ditches; can be quite weedy in ditches and wet meadows; from Newfoundland to Alaska, south to California, South Dakota and Virginia.

NOTES: Lady fern is often confused with the **wood ferns** (*Dryopteris* **spp.**, p. 200), but those species have rounder spore clusters (sori). • The species name *filix-femina*, 'lady fern,' was taken from a myth that this fern would sneak around at night in search of male fern (*Dryopteris filix-mas*).

OSTRICH FERN
MATTEUCCIE FOUGÈRE-À-L'AUTRICHE
Matteuccia struthiopteris

GENERAL: Erect, 0.5–2 m tall, perennial fern; stalks smooth; often forming large patches, from black, underground stolons and dark brown, scaly rhizomes.

LEAVES: In whorls, of 2 distinct types (fertile and sterile); sterile leaves **ostrich-plume-like**, widest in upper ⅓, rapidly tapering to pointed tip and gradually tapering to base, **up to 2 m long and 35 cm wide**, 1–2 times pinnately divided, with veins of leaflets not forked; fertile leaves described with spore clusters.

SPORE CLUSTERS: Covered by down-rolled edges of fertile leaves; fertile leaves up to 60 cm tall, club-shaped, rigid, olive green at first, **turning dark brown**; mid-summer.

WHERE FOUND: In hardwood swamps and on streambanks; from Alaska to Newfoundland, south to South Dakota, Missouri and Virginia.

NOTES: Ostrich fern is the largest fern in our region. Its sterile leaves wither with the first frost, but the dark brown fertile leaves remain upright through the winter. • The young, coiled leaves (fiddleheads) of ostrich fern are harvested commercially in eastern Canada, and then sold as a frozen vegetable. **Caution:** The fiddleheads of some ferns, particularly those of **bracken fern** (***Pteridium aquilinum***), are **poisonous**. • The genus *Matteuccia* was named for the Italian physicist Carlo Matteuci, who lived from 1800 to 1868.

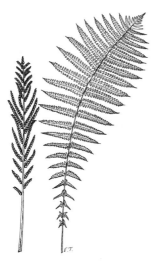

SPINULOSE WOOD FERN
DRYOPTÈRE SPINULEUSE • *Dryopteris carthusiana*

GENERAL: Perennial fern; from stout rhizomes.

LEAVES: Up to 70 cm long, broad, yellowish green, glandless, 2–3 times pinnately divided, not very lacy, asymetrical; smallest leaflets (pinnules) with delicate teeth; stalks long, with many brown scales; sub-leaflets (pinnules) on lower (basal) side of largest leaflets (pinnae) longer than their adjacent sub-leaflets; **stalks very chaffy, with brown papery scales**.

SPORE CLUSTERS: Rounded dots (sori) on underside of leaves, covered by a flap of delicate tissue (indusium); summer.

WHERE FOUND: In swamps, moist forests, wet meadows, ditches and thickets and on streambanks; from Newfoundland to Alaska, south to California, Iowa and North Carolina.

NOTES: This species was called **D. spinulosa** in earlier literature. • Two other wetland species are also fairly common in habitats similar to those of spinulose wood fern. • **Crested wood fern** (**D. cristata**) has glossy, narrow, less-dissected leaves, with broadly triangular lower leaflets (pinnae), on which the lowermost sub-leaflets (pinnules) are not stalked. Its range extends from southern British Columbia to Alberta to Newfoundland, south to Idaho, Louisiana and North Carolina. • **Evergreen wood fern** (**D. intermedia**, previously included in **D. austraica** or **D. spinulosa**) has broad, very lacy leaves that remain dark green long into the winter, and that have glandular hairs on the thin membranes (indusia) covering their spore dots. The sub-leaflets (pinnules) on the lower (basal) side of the largest leaflets (pinnae) are shorter than their adjacent sub-leaflets, and they are stalked. It is an eastern North American species. • Several other species of wood fern are found in upland habitats. • Spinulose wood fern is eaten in small quantities by rabbits, hares and deer. • The genus name *Dryopteris* was derived from the Latin *dryo*, 'oak,' and *pteris,* 'fern.'

MARSH FERN
THÉLYPTÈRE DES MARAIS • *Thelypteris palustris*

GENERAL: Slightly hairy, perennial fern, 10–70 cm tall; from **long black rhizomes.**

LEAVES: Tapering from middle to tip, but not to base, 20–60 cm long, 10–15 cm wide, minutely hairy, **twice pinnately divided**; smallest leaflets (pinnules) rounded at tip, thick, rolled under at edge (especially in fertile leaves); **veins forked**.

SPORE CLUSTERS: Rounded, crowded **spore dots (sori)** on underside of leaflets (pinnules) midway between edge and midvein; mid- to late summer.

WHERE FOUND: In swamps, marshes, wet ditches and moist forests and on lakeshores, riverbanks and streambanks; from southeastern Manitoba to Newfoundland, south to Oklahoma and Georgia.

NOTES: This species was previously called **Dryopteris thelypteris**. • **New York fern** (**T. noveboracensis**) has leaves that taper to the base. It grows in swamps or uplands from Minnesota to Newfoundland, south to Arkansas and Georgia. • Early writers suggested that these leaflets, with their down-rolled edges, reminded them of snuff boxes. • The genus name *Thelypteris* means 'female fern'; *palustris* means 'of the marsh or swamp.'

CINNAMON FERN
OSMONDE CANNELLE • *Osmunda cinnamomea*

GENERAL: Showy, green, perennial, fern **80–160 cm tall**, in rounded clumps from thick rhizomes (often covered with stubble of old leaf bases) and fibrous roots.

LEAVES: Of 2 types (sterile and fertile), with **clumps** of sterile leaves around 1 or more fertile leaves; sterile leaves once pinnately divided into lobed leaflets (pinnae), with tufts of rusty wool at base; fertile blades described with spore clusters.

SPORE CLUSTERS: Cinnamon-coloured, naked, not covered by leaf edges or membranes (indusia), forming large, rounded clumps and covering fertile leaves; fertile leaves **erect, smaller** than sterile leaves, with contracted, contorted blades and **grooved stalks**; late summer.

WHERE FOUND: In swamps, marshes, moist forests, ditches, wet fields and wet shrubby thickets and on streambanks; from Ontario to Newfoundland, south to New Mexico, Texas and Florida.

NOTES: **Royal fern** (*O. regalis*) has leaves that are twice divided and up to 180 cm tall, with fertile leaflets produced at the tip of normal blades. It grows on lakeshores and peatland edges and in acidic woodlands from Ontario to Newfoundland, south to Mexico and Florida. • **Interrupted fern** (*O. claytoniana*) has once-divided leaves with rounded leaflets (pinnae). It produces 3–5 pairs of dark brown, contracted, fertile leaflets near the middle of some leaves. Interrupted fern grows in moist woods from southeastern Manitoba to Newfoundland, south to Arkansas and Georgia. • Pioneers believed that 1 bite into the unfolding leaves of cinnamon fern prevented them from getting a toothache for a full year! Royal fern was once believed to have the power to heal wounds and broken bones, and to give 'eternal life' to anyone who drank the sap that 'floweth from the stem.' Its coiled leaves (fiddleheads) are sometimes eaten as a delicacy. They have an acrid, nutty flavour, and they are not nearly as good as the fiddleheads of **ostrich fern** (*Matteuccia struthiopteris*, p. 194). **Caution:** The fiddleheads of some ferns, particularly those of **bracken fern** (*Pteridium aquilinum*), are **poisonous**.

O. regalis

O. claytoniana

201

BRYOPHYTES

This section includes some common mosses and liverworts found in wetlands. These plants are distinguished from higher plants by their small size, their frequent lack of vascular (water conducting) tissue and their production of spores (rather than seeds). The sporophyte remains permanently attached to the gametophyte throughout the life cycle, and the spores are produced in capsules or occasionally in specialized chambers (in some liverworts). Mosses and liverworts are found in a wide range of habitats, but some require specific conditions, and these can be helpful when determining ecosystem or habitat types.

To identify bryophytes, you should be familiar with both their life cycle and their morphology/anatomy. The small green plants (gametophytes) gain energy from photosynthesis, and absorb nutrients and moisture from the atmosphere. The stalked capsules full of spores (sporophytes) depend on the gametophyte for their nutrition. Both parts of the plant can be helpful for identification, but sporophytes are not always present. Often the best time to collect and identify these plants is in the spring and autumn—when there are no biting insects!

Peat mosses (*Sphagnum* spp., pp. 218–22) are treated separately because of their uniqueness, abundance, value as a habitat indicator and identification problems.

The following illustrations will help you to identify some of the mosses and liverworts discussed in this guide.

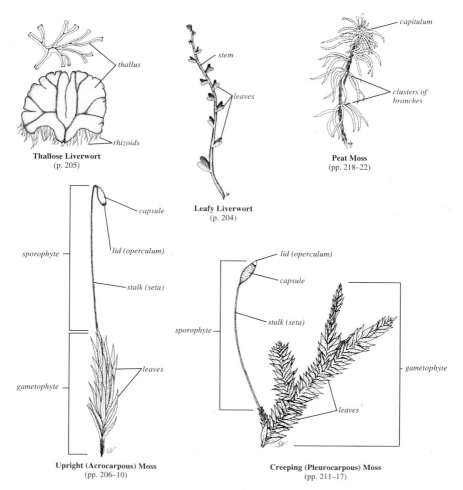

Thallose Liverwort
(p. 205)

Leafy Liverwort
(p. 204)

Peat Moss
(pp. 218–22)

Upright (Acrocarpous) Moss
(pp. 206–10)

Creeping (Pleurocarpous) Moss
(pp. 211–17)

203

KEY TO THE BRYOPHYTES

Plants ribbon-like or leafy; leaves in 2 rows, often lobed; capsules splitting in 4;
Liverworts

Plants always leafy; leaves not lobed, usually spirally arranged on stem; capsules opening by a lid (operculum);
Mosses

Plants with leaves and stems

Plants with flattened, ribbon-like bodies, lacking leaves

Branches in clusters, usually forming enlarged heads at stem tips; leaves with narrow, green cells forming a net-like pattern, interspersed with large, clear cells

Branches single or absent; leaves composed of similar, green cells

Plants upright, often forming tufts of cushions; spore capsules produced at stem tips

Plants creeping or trailing, usually forming mats; spore capsules produced from stem sides

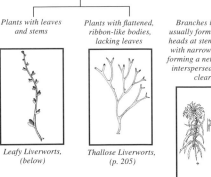

Leafy Liverworts, (below)

Thallose Liverworts, (p. 205)

Peat Mosses, (pp. 218–222)

Acrocarpous Mosses, (pp. 206–210)

Pleurocarpous Mosses, (pp. 211–217)

FLOATING BOG LIVERWORT
Cladopodiella fluitans

GENERAL: Dark green, leafy liverwort, 1.5–3 cm long, with dark stems.

LEAVES: Attached to stem on an angle, sparse (not usually overlapping), heart-shaped; small underleaves usually found near stem tip.

SPOROPHYTES: From a pouch-like sheath of fused leaves (perianth), with a clear stalk and a capsule that splits into 4 parts to release spores, living only 1–2 days.

WHERE FOUND: In wet microhabitats in bogs, fens or conifer swamps, usually on peat moss hummocks; in eastern North America and reported from British Columbia and Washington.

NOTES: Floating bog liverwort's small size and inconspicuous habitat make it elusive, but persistence and patience will reveal its presence in many of our bogs and fens. • Liverworts are home to many different types of microscopic freshwater organisms. • The genus name *Cladopodiella* means 'foot-like branch,' and it refers to the short branches on which the female reproductive structure (archegonia) form. The species name *fluitans* simply means 'floating.'

FLOATING SLENDER LIVERWORT
Riccia fluitans

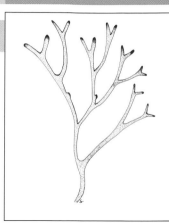

GENERAL: Flat, ribbon-like liverwort, green, translucent, **repeatedly divided into segments less than 1 mm wide**, floating just below water surface; rhizoids only in stranded (terrestrial) plants.

LEAVES: Absent

SPOROPHYTES: Reduced to minute capsules, in air chambers in ribbon-like segments.

WHERE FOUND: Stranded on muddy shores or floating just below surface of ponds, lakes or slow streams and rivers; throughout North America above the tropics.

NOTES: Several other *Riccia* species grow in northern Ontario wetlands, but they are rare, and they do not have branched, slender, ribbon-like segments. • Floating slender liverwort can be confused with **purple-fringed liverwort** (*Ricciocarpos natans*, below), but that liverwort is wider, and it has dense, root-like rhizoids. Purple-fringed liverwort is usually found in the spring and autumn, when it is not out-competed by more vigorous plants. • Floating slender liverwort is a colonist plant, invading muddy shores and tire tracks. Plants are often carried to new habitats on the shells of turtles or on the feet and feathers of ducks. When it dries, the plant curls up into a ball. • The species name *fluitans* means 'floating.'

PURPLE-FRINGED LIVERWORT
Ricciocarpos natans

GENERAL: Silvery-green liverwort, flattened, **fan-shaped**; segments branched, containing large air chambers, less than 5 mm wide, with a distinct **groove** on upper surface, and purplish scales plus dense, root-like rhizoids beneath; **floating in mats** on water surface.

LEAVES: Absent.

SPOROPHYTES: Reduced to capsules embedded in plant body, visible as black mass just under groove on upper surface; early summer.

WHERE FOUND: On surface of quiet pools in streams, swamps, ponds and lakes; occasionally stranded on muddy shores; throughout North America, south of the Arctic.

NOTES: Stranded plants can be confused with **floating slender liverwort** (*Riccia fluitans*, above), but the rhizoids, purple-scaly undersurface and grooved upper surfaces of purple-fringed liverwort are distinctive. • Many small aquatic animals and insects live among these plants. Purple-fringed liverwort also provides food for waterfowl, turtles and fish. • The genus name *Ricciocarpos* refers to the similarity between this genus and the genus *Riccia*. The species name *natans* means 'swimming.'

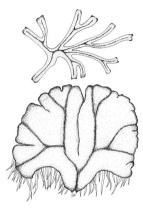

205

KEY TO ACROCARPOUS MOSSES

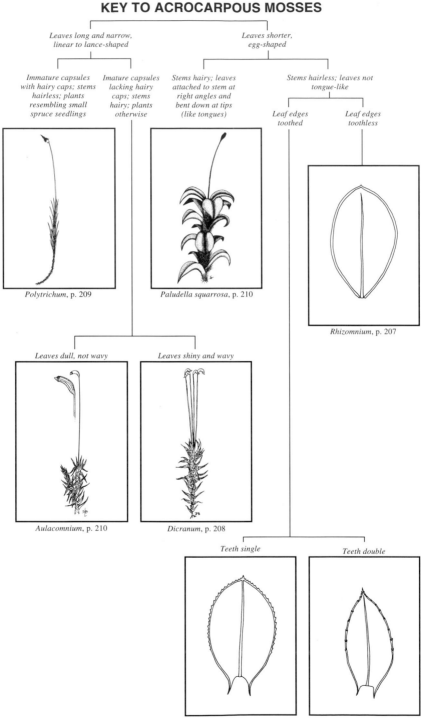

Leaves long and narrow, linear to lance-shaped

Leaves shorter, egg-shaped

Immature capsules with hairy caps; stems hairless; plants resembling small spruce seedlings

Imature capsules lacking hairy caps; stems hairy; plants otherwise

Stems hairy; leaves attached to stem at right angles and bent down at tips (like tongues)

Stems hairless; leaves not tongue-like

Leaf edges toothed

Leaf edges toothless

Polytrichum, p. 209

Paludella squarrosa, p. 210

Rhizomnium, p. 207

Leaves dull, not wavy

Leaves shiny and wavy

Aulacomnium, p. 210

Dicranum, p. 208

Teeth single

Teeth double

Plagiomnium, p. 207

Mnium, p. 207

MNIUMS
MNIE, PLAGIOMNIE, RHIZOMNIE
Mnium spp., *Plagiomnium* spp., *Rhizomnium* spp.

GENERAL: Erect, leafy mosses, often quite large, dull green or yellow-green; stems hairy, producing **sterile, horizontal branches**; some forming large patches.

LEAVES: Rounded to egg-shaped, with a **strong midrib**; edges toothless or with single or double teeth (depending on genus).

SPOROPHYTES: From stem tips, clustered or single; **capsules usually nodding.**

WHERE FOUND: In swamps, moist forests and fens and on shaded streambanks and cliffs soaked with ground water; also in upland forests; across North America.

NOTES: The leaves of some mniums become very twisted and wrinkled when dried, and they need to be rehydrated before species can be identified. Soaking a clump in a stream for a couple of minutes works quite well. • Several species in this group are common in Ontario. The mniums can be separated into the following groups using a hand lens.

Mniums whose leaves lack teeth and have a thick border (*Rhizomnium* spp.): Pointed round moss (*R. punctatum*) is often red-tinged, and its midrib extends slightly beyond the leaf tip as small bump. In Ontario, it is very common in cedar swamps. Generally, pointed round moss grows on rock, soil, humus or rooting wood in woodlands, from Ontario to Newfoundland, south to Arkansas and Georgia. • **Felt round moss (*R. pseudopunctatum*)** is dark green, and its midrib ends below the leaf tip. It grows in fens, streams and cedar swamps, from Alaska to Greenland, south to Colorado, Wisconsin and Maine.

Mniums with single-toothed leaves (*Plagiomnium* spp.): Common leafy moss (*P. medium*) has leaves that are toothed from the tip to the base, and it has stalked capsules that are often clustered. It grows in conifer swamps and wet woodland sites, from Alaska to Greenland, south to California and Tennessee. • **Woodsy mnium (*P. cuspidatum*)** has leaves that are toothed only on their upper half, and it has single, stalked capsules. It grows in moist forests, from Alaska to Labrador, south to Arizona, Colorado, Alabama and Florida.

Mniums with double-toothed leaves (*Mnium* spp.): Edged lantern moss (*M. marginatum*) has solitary stalked capsules and reddish stems. It grows along streambanks and sometimes on cliffs and bluffs, from Alaska to Labrador, south to Arizona and North Carolina.

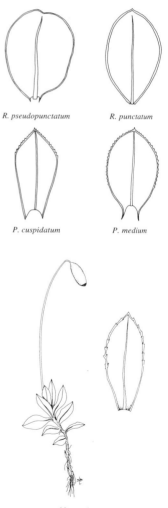

R. pseudopunctatum   *R. punctatum*

P. cuspidatum   *P. medium*

M. marginatum

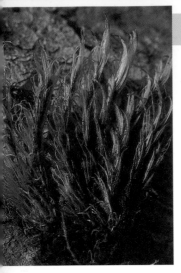

WAVY MOSS
Dicranum undulatum

GENERAL: Robust, densely tufted moss; stems reddish brown woolly, 2–15 cm tall.

LEAVES: Compact, not very shiny, **5.5–9 mm long, narrow, wavy, slightly curved, toothed, with a single midrib** extending to a broadly pointed leaf tip.

SPOROPHYTES: Single, from stem tips, long-stalked; capsules 2.5–4 mm long, curved, with a **long-pointed lid (operculum)**; **empty capsules furrowed when dry** and persist throughout year.

WHERE FOUND: On peat moss hummocks in bogs, fens and conifer swamps; from Alaska to Greenland, south to the northern U.S.

NOTES: Another species of this genus, **D. polysetum** (also called **wavy moss**), closely resembles *D. undulatum*. It is distinguished by having many (rather than 1) spore capsules on each plant, and its leaves are shinier, more wide-spreading and more narrowly lance-shaped with longer, slender, pointed tips, when compared to *D. undulatum* leaves. *D. polysetum* usually grows on humus (occasionally on mineral soil) in higher areas, such as forests and woodlands, from Alaska to Newfoundland, south to Washington, Missouri and North Carolina. It can also be found in cedar or deciduous swamps, but it does not grow in wet, acidic sites, such as bogs and black spruce swamps. • Most other species of *Dicranum* lack wavy leaves, and they are typically found in drier habitats. • **Broom moss (D. fuscescens)** is a common, medium-sized *Dicranum* that is often difficult to identify with certainty. It has leaves that are swept to 1 side and that are distinctly keeled (V-shaped) rather than rounded near their tips. Broom moss grows on wood, usually at the base of trees or on decaying wood, in swamps or upland forests from Alaska to Greenland and Newfoundland, south to California, Colorado and Tennessee. • The genus name *Dicranum* means '2 teeth,' referring to the pairs of teeth that surround the opening of the capsule. These teeth control spore dispersal by opening and closing with changes in humidity. The species name *polysetum* means 'many-stalked,' referring to the several stalked spore capsules at the tip of a stem.

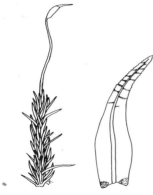

D. undulatum

D. polysetum

D. polysetum

HAIRCAP MOSSES
POLYTRIC • *Polytrichum* spp.

GENERAL: Erect, robust, tufted or mat-forming mosses, resembling **little spruce trees**; stems rigid, 1–45 cm tall.

LEAVES: Stiff, 6–10 mm long, lance-shaped, pointed, sometimes with a long, slender bristle-tip (awn); midribs single, thick, with dense rows of green cells (lamellae) on upper surface; edges finely toothed or toothless, often rolled upward and overlapping over midrib, giving leaves a needle-like appearance; male plants often have distinctive cup-like rosettes of perigonial leaves at tips of their stems.

P. commune

SPOROPHYTES: From stem tips, solitary; capsules 4-sided, covered by a hairy cap (calyptra) when young; produced whenever conditions are favourable.

WHERE FOUND: From bogs, fens and other wet areas to dry, sandy or gravelly sites; on soil or humus; across Canada and the U.S.

NOTES: Common haircap moss (*P. commune*) is 4–45 cm tall, and it has saw-toothed leaf edges and 3–5 mm long capsules that are deeply constricted at the base. Common haircap moss is often over 10 cm tall, and it is the only haircap moss to grow over 20 cm tall. It is a very common, cosmopolitan species that grows on soil or humus in a wide range of wet conditions. • Common haircap moss could be confused with *P. formosum*, which also grows in moist forests and swamps and along bog edges, but *P. formosum* is less common, and its capsules are not deeply constricted at the base. It grows in woodlands and beside streams, from Alaska to Newfoundland, south to Washington, Wyoming and North Carolina. • Bog haircap moss (*P. strictum*) is 4–16 cm tall, with

P. strictum

inconspicuously awned, toothless leaves, that have their edges folded up over the midvein. Its capsules are constricted at the base, and its stems are matted together by a woolly mass of rhizoids. Bog haircap moss grows on hummocks in and around bogs from Alaska to Greenland, south to North Carolina, Minnesota, Colorado and Washington. • **Juniper haircap moss** (*P. juniperinum*) is very similar to bog haircap moss, and these 2 species are easily confused. Juniper haircap moss grows on dry, upland sites, however, and its stems lack the dense, white rhizoids found in bog haircap moss. Juniper haircap is found from Alaska to Greenland, south to California, Colorado and Alabama. • Small mammals, such as mice and moles, eat *Polytrichum* plants (gametophyes), and the stalked capsules are eaten by grouse and used by various birds for nesting material. • The genus name *Polytrichum* means 'many-hairs,' and it refers to the hairy caps on the young spore capsules.

TONGUE MOSS
Paludella squarrosa

GENERAL: Robust moss, in dense tufts 4–7 cm tall; **stems with purplish hairs.**

LEAVES: In 5 vertical rows (ranks) on stem; **spreading at right angles, abruptly curved downwards** at tip (like someone sticking their tongue out), **strongly keeled** (V-shaped), with a **single, slender midrib.**

SPOROPHYTES: Stalks 2.5–4 cm long, reddish brown; **capsule slightly curved, 2–3 mm long, brown.**

WHERE FOUND: In calcareous fens and occasionally around sedgy borders of beach pools; from Alaska to Greenland, south to Montana and New York.

NOTES: Tongue moss is not easily confused with any other moss. This distinctive moss is a treat to find, and it is a sure indicator of fen-like conditions. • The genus name *Paludella* means 'little marsh,' and *squarrosa* refers to the way the leaves spread at right angles.

RIBBED BOG MOSS
AULACOMMIE DES MARAIS
Aulacomnium palustre

GENERAL: Leafy, **bright yellow-green**, tufted moss, erect, 3–9 cm tall, robust, **branched; stems covered with reddish-brown hairs;** sterile plants sometimes have starry-looking gemmae at tips.

LEAVES: Oblong-lance-shaped, pointed, 2–4 mm long with a **distinct midrib,** strongly crisped and **contorted when dry.**

SPOROPHYTES: From stem tips; stalks smooth; **capsules cylindrical and curved.**

WHERE FOUND: In bogs, fens and conifer swamps, often forming tufts in peat moss hummocks, or large patches on cliffs soaked with ground water; from Alaska to Greenland, south to California, Colorado and Louisiana.

NOTES: Wavy mosses (*Dicranum* **spp.**, p. 208) are similar to ribbed bog moss, but they are darker in colour, their stems do not branch, and they lack stalked gemmae. • Ribbed bog moss is a pioneering species in bogs that have been burned over. • The species name *palustre* means 'of the marsh or swamp.'

KEY TO PLEUROCARPOUS MOSSES

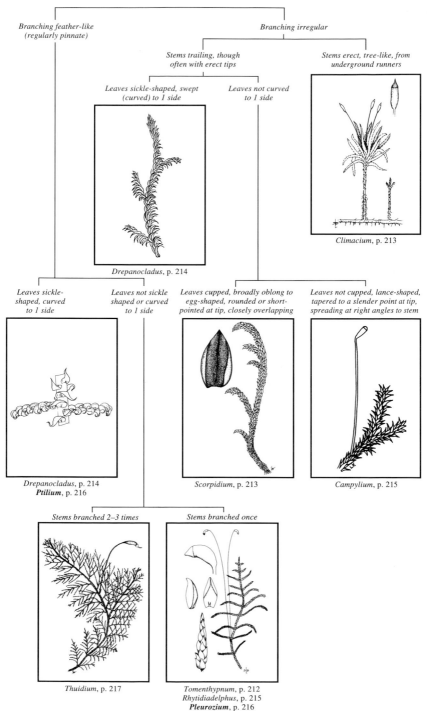

Branching feather-like (regularly pinnate)

Branching irregular

Stems trailing, though often with erect tips

Stems erect, tree-like, from underground runners

Leaves sickle-shaped, swept (curved) to 1 side

Leaves not curved to 1 side

Climacium, p. 213

Drepanocladus, p. 214

Leaves sickle-shaped, curved to 1 side

Leaves not sickle shaped or curved to 1 side

Leaves cupped, broadly oblong to egg-shaped, rounded or short-pointed at tip, closely overlapping

Leaves not cupped, lance-shaped, tapered to a slender point at tip, spreading at right angles to stem

Drepanocladus, p. 214
Ptilium, p. 216

Scorpidium, p. 213

Campylium, p. 215

Stems branched 2–3 times

Stems branched once

Thuidium, p. 217

Tomenthypnum, p. 212
Rhytidiadelphus, p. 215
Pleurozium, p. 216

211

FUZZY BROWN MOSS
Tomenthypnum nitens

GENERAL: Erect, branched moss, in **golden-brown tufts** up to 15 cm tall; stems with conspicuous, **brown, fuzzy hair; branches short, straight, wide-spreading.**

LEAVES: Erect-spreading, slender-pointed, 3–4 mm long, **strongly pleated, lustrous when dry;** midrib thin, ending below leaf tip.

SPOROPHYTES: Solitary, from stem sides; stalks slender, about 5 cm long, reddish; capsules less than 3 mm long.

WHERE FOUND: In rich fens and calcareous conifer swamps; fen indicator; from Alaska to Greenland, south to Washington, Colorado, Minnesota and New Jersey.

NOTES: Fuzzy brown moss is common in this region, but it is often overlooked because it blends in with brownish ground cover. • A similar species, **river moss** (*Brachythecium rivulare*), lacks the conspicuous, fuzzy, brown hairs. It grows in and beside brooks and waterfalls from Alaska to Labrador, south to Arizona, Arkansas and North Carolina. • **Clay pigtail moss** (*Hypnum lindbergii*) and other hypnums are similar in appearance, as the genus name *Tomenthypnum* suggests, but the leaves of *Hypnum* species do not have a midrib, and the plants typically grow in swampy habitats. • The first half of the genus name *Tomenthypnum* refers to the brown woolly hair (tomentum) on the stems. The species name *nitens* means 'shiny,' and it refers to the lustrous leaves.

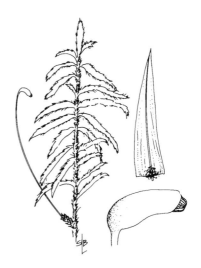

TREE MOSS
CLIMACIE ARBUSTIVE • *Climacium dendroides*

GENERAL: Tree-like moss, with soft branches from near the tips of erect, vertical shoots, **3–9 cm tall**; from **underground, horizontal stems**.

LEAVES: Yellowish-green, 2–3 mm long, **rounded at tip, densely overlapping**.

SPOROPHYTES: From side of upper branches, rarely produced, but can grow from any branch; capsules 1.5–3 mm long, cylindrical, with a beaked lid (operculum); stalks 18–45 mm long.

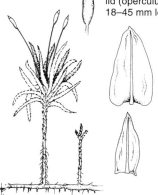

WHERE FOUND: In damp thickets, hardwood and conifer swamps and moist forests and on streambanks and lakeshores; from Alaska to Greenland, south to Virginia, Illinois and California.

NOTES: Another moss with a tree-like form, but with nodding branches (some people suggest it looks like a 'stepped-on' tree moss), is ***Thamnobryum alleghaniensis***. This moss grows in swamps and on wet rocks, from Ontario to Nova Scotia, south to Arkansas and Georgia. • The species name *dendroides* means 'tree-like,' and it refers to the growth form of this moss.

SCORPION'S TAIL
Scorpidium scorpioides

GENERAL: Shiny, usually dark brown (occasionally yellow-green or red-gold), **worm-like moss, often slimy**; **stems erect, curved at tips**, up to 25 cm long, with few branches.

LEAVES: Elliptic to egg-shaped, usually 2–4 mm long, with a **short, double midrib**, cupped, often hooded, crowded, curved to 1 side (as if windblown) at branch tips.

SPOROPHYTES: From stem sides, solitary; **stalks (setae)** up to 5 cm long; **capsules cylindrical, slightly curved, less than 3 mm long.**

WHERE FOUND: In rich fens, **usually submerged in shallow, stagnant water**; from Alaska to Greenland, south to New Jersey, Ohio and Montana.

NOTES: Many people walk all over scorpion's tail, looking for larger plants that are fen indicators. This moss is usually submerged in shallow water, and it may be covered with limey deposits. • Scorpion's tail is distinguished from **sickle mosses (*Drepanocladus* spp.**, p. 214) by its robust, worm-like (julaceous) appearance, its short, double midribs and its rounded to short-pointed leaf tips. • Both parts of the scientific name, *Scorpidium* and *scorpioides*, mean 'like a scorpion's tail,' and they refer to the curved, pointed branch tips.

213

SICKLE MOSSES, CURVED BRANCH MOSSES
Drepanocladus spp., *Hematocaulis* spp., *Limprichtia* spp., *Sanionia* spp.

GENERAL: Robust, wetland mosses; **stems thin, creeping, freely branched and slightly erect**; forming extensive mats.

LEAVES: Lance-shaped, with long, slender tips, **curved to 1 side** (as if windblown), with a single midrib not reaching the leaf tip.

SPOROPHYTES: From stem sides, solitary, **long-stalked**; capsules curved.

WHERE FOUND: Most are aquatic, growing in swamps, fens, rivers, streams, wet fields or ditches, or on lakeshores; across North America.

NOTES: Identification to the species level requires microscopic examinination the leaf cells. The leaves of **hook moss** (*Sanionia uncinata*, previously known as *D. uncinatus*) have distinct midveins and hooked tips, and they are deeply pleated. Hook moss is found in drier situations than other species of this genus, from Alaska to Greenland, south to Mexico, Minnesota and Pennsylvania. It often forms large, pure mats at the moist bases of large boulders, where it is associated with **stair-step moss** (*Hylocomium splendens*, p. 217). • Sickle moss could be confused with **scorpion's tail** (*Scorpidium scorpioides*, p. 213). • *Hematocaulis vernicosus* (previously known as *D. vernicosus*) and *D. aduncus* are good fen indicators, whereas *L. revolvens* (previously known as *D. revolvens*) is usually found in poor fens. • These mosses provide habitat for many small aquatic animals, which in turn provide food for waterfowl, amphibians, reptiles and fish. • *Drepanocladus* means 'curved branch,' and it refers to the curved leaves, which give the branch tips a curved, sickle-like appearance.

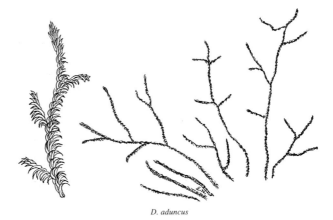

D. aduncus

STARRY CAMPYLIUM
Campylium stellatum

GENERAL: **Golden brown moss**; **stems erect, giving tufted appearance**, often **forming shining mats.**

LEAVES: Crowded and spreading, giving plant a starry appearance; 1.5–3 mm long, broadly oval at base, slender tipped; midrib short or lacking, single or forked.

SPOROPHYTES: From stem sides, long-stalked; **capsules curved**, about 3 mm long; most common in autumn and spring, but present all year.

WHERE FOUND: In fens; a fen indicator; from Alaska to Greenland, south to Washington, New Mexico, Minnesota and North Carolina.

NOTES: Several other species of *Campylium* are found on wet logs, soil and rocks in moist (usually deciduous) forests in this region. *C. polygamum* has leaves with a strong, single midrib, and it grows in bogs, swamps or fens from Alaska to Greenland, south to California, Colorado and Florida. • **Giant water moss** (*Calliergon giganteum*), **Richardson's water moss** (*Calliergon richardsonii*) and **straw-coloured water moss** (*Calliergon stramineum*) could be confused with starry campylium, but these water mosses are typically found in fens, although they may also grow in swamps, along creeks, in bog pools and in 'moats' surrounding peatlands. • Positive identification of *Campylium* species requires microscopic techniques, but if you find a moss like this, growing in golden-yellow mats, you are probably standing in a fen and looking at starry campylium. • The genus name *Campylium* means 'bent over,' and it refers to the capsules, which have a distinctive curve. The species name *stellatum* means 'starry.'

SHAGGY MOSS, ELECTRIFIED CAT'S TAIL MOSS
HYPNE TRIQUÈTRE • *Rhytidiadelphus triquetrus*

GENERAL: Very coarse, shaggy moss, dark green to yellow-green; branches wide-spreading; **stems red or orange**, creeping to ascending, up to 20 cm long, forming loose tufts.

LEAVES: Crowded (especially toward stem tips), 1.5–5 mm long, **delicately pleated, wrinkled near tip**, clasping at base, erect to wide-spreading; midrib double, extending just past middle of leaf.

SPOROPHYTES: From stem sides, scattered solitary; stalks 15–45 mm long, **reddish**; capsules 1.5–3 mm long, **reddish brown, curved, nodding**.

WHERE FOUND: In rich conifer swamps and moist upland forests; on soil, humus and rotting logs; from Alaska to Labrador, south to California, Montana, Arkansas and Florida.

NOTES: The wide-spreading branches and pleated, wrinkled, clasping leaves distinguish shaggy moss from other species. • The genus name *Rhytidiadelphus* means 'with wrinkled leaves.' The species name *triquetrus* refers to the leaves, which are in 3 vertical rows (ranks).

215

SCHREBER'S MOSS, BIG RED STEM
HYPNE DE SCHREBER • *Pleurozium schreberi*

GENERAL: Robust, **light green or yellowish feathermoss**, 7–16 cm tall; **stems bright red**, creeping, **pinnately branched** (like a feather); often forming **extensive mats**.

LEAVES: Egg-shaped, **concave**, with a short, double midrib, **loosely overlapping**, giving branches a succulent appearance; **tips slightly curved outward**, rounded, but appearing pointed because of incurved edges.

SPOROPHYTES: From stem sides, solitary; stalks red; capsules reddish brown, cylindrical, curved.

WHERE FOUND: In woods and occasionally in bogs (often forming patches on drier, shaded hummocks) and swamps, on soil, humus, logs and rock; from Alaska to Greenland, south to Oregon, Arkansas and North Carolina.

NOTES: Schreber's moss is one of our most common and distinctive mosses. A simple test to identify Schreber's moss is to hold the plant by its tip while pulling down to removing its leaves and branches, revealing a bright red stem. • Schreber's moss often grows with **stair-step moss** (*Hylocomium splendens*, p. 217), but that moss has fern-like branches that form horizontal 'steps' up the stem. • The genus name *Pleurozium* means 'ribbed,' and it refers to the regular, rib-like arrangement of branches on the stem.

PLUME MOSS
HYPNE PLUMEUSE • *Ptilium crista-castrensis*

GENERAL: Robust, bright green to golden feathermoss; stems up to 10 cm long, regularly pinnately branched, feather-like, often trailing, but upper $^1/_3$ usually erect; branches progressively smaller towards tip; often forming large, dense mats.

LEAVES: Shiny, bright green, pleated, delicate, with a short double midrib; **branch leaves** hooked at tips, **curved to 1 side (as if windblown), smaller than stem leaves.**

SPOROPHYTES: From stem sides, solitary; stalks reddish; capsules cylindrical, brown.

WHERE FOUND: In swamps (with peat moss and Schreber's moss) and moist upland forests, on logs, humus, rock and soil; from Alaska to Labrador, south to Idaho, Iowa and North Carolina.

NOTES: Plume moss frequently grows with **Schreber's moss** (*Pleurozium schreberi*, above) and **stair-step moss** (*Hylocomium splendens*, p. 217). Schreber's moss is less feathery, with less regular branching, and it has distinctive red stems. Stair-step moss has branches that are pinnately divided 2–3 times, and that form successive layers with each year's of growth. • *Ptilium* means 'feather,' and *crista-castrensis* means 'a soldier's plume,' in reference to the feather-like appearance of this beautiful moss.

COMMON FERN MOSS
MOUSSE FOUGÈRE • *Thuidium delicatulum*

GENERAL: Rather robust, **yellow-green moss, 2–3-times pinnately branched, with a delicate, fern-like appearance**; stems 3–8 cm long.

LEAVES: Stem leaves less than 1.5 mm long, sparse; **branch leaves less than 0.5 mm long**.

SPOROPHYTES: From stem sides, solitary; **stalks reddish**, 5 cm long; capsules slightly curved, less than 4 mm long, with a **long-pointed lid** (operculum).

WHERE FOUND: In cedar swamps and moist forests on soil, humus, rock, tree roots, tree trunks and rotten logs; an indicator of rich, moist conditions; from Alaska to Labrador, south to Texas and Florida.

NOTES: No other mosses in this region have branching patterns and reduced leaves like common fern moss. • **Stair-step moss** (*Hylocomium splendens*) is a large, common feathermoss with fern-like branches that are 2–3 times pinnately divided. It has a distinctive growth pattern that produces horizontal 'steps' up the stem, with the production of a flattened, spreading branch (frond) each year. Stair-step moss is one of the most common feathermosses in Canadian forests. It grows in moist to wet forests on a wide range of substrates, from Alaska to Greenland, south to Oregon, Colorado and North Carolina. • The genus name *Thuidium* was taken from the Latin *thuja*, 'cedar,' because these mosses were thought to resemble the branches of cedar trees. The species name *delicatulum* refers to its delicate form.

T. delicatulum

Hylocomium splendens

Hylocomium splendens

Hylocomium splendens

217

Peat mosses • *Sphagnum* spp.

There are over 30 peat mosses (*Sphagnum* spp.) in Ontario. Some can be recognized in the field using growth form, colour, leaves, branches and habitat. The treatment presented here is very basic, and it describes 10 common peat mosses in 5 major groups using field characteristics.

For hundreds of years *Sphagnum* species were labelled by botanists as 'difficult to identify,' but the major sections of this large genus can be recognized using features such as growth form, colour, branching patterns, habitat and especially leaf characteristics. Some species require microscopic examination of leaf and stem cellular features for identification. See Crum's *Mosses of the Great Lakes Forest* (1983) and *A Focus on Peatlands and Peat Mosses* (1992) for a more detailed treatment of peat mosses.

Peat mosses are **robust, erect mosses** of moist habitats, often forming **large, rounded hummocks**. Their **branches grow in clusters (fascicles)**, spirally arranged on stem and **crowded in a head-like tuft** (capitulum) **at the stem tip**. Peat mosses have 2 kinds of leaves the branch and stem, which are usually very different. Leaf size and shape are useful in species identification.

Peat moss sporophytes grow from stem tips, elevated on stalks (setae). Their **capsules are round and usually black**. Mature capsules dry and constrict until their lids are blown off with an audible pop, sending spores a few centimetres away from the plant. The pressure inside these tiny capsules at maturity is 4–6 atmospheres (similar to the pressure in truck tires). Perhaps some day, sitting in a bog and taking in the warm summer sun, you will hear peat moss capsules popping off their lids.

Peat mosses grow in bogs, fens and swamps throughout North America. Some species are habitat specific and make excellent indicators of different wetland types. The ecology of peat mosses is very important to northern Ontario. Peat moss regulates drainage systems by impeding drainage and storing water. In basic conditions, these mosses use ion exchange to create an acidic habitat for themselves and other acid-loving plants. They also provide seed beds for many trees. Apparently, peat mosses are rarely eaten by wildlife, although they have been mixed with feed for livestock.

Peat moss has been used extensively throughout history. Native people discovered that these plants have very good absorptive and antiseptic properties. They used peat moss for dressings on wounds, diapers (their babies rarely suffered from diaper rash), menstrual pads, bedding and clothing. Pioneers used peat moss in bedding, medicinal baths, insulation, lantern wicks, dyes and diapers and as a fuel for heating. During the Napoleonic and First World wars, peat moss was used extensively to make dressings, which were 3–4 times more absorbent than cotton. Many of the leaf cells are large, empty holding chambers for water, which allow these plants to absorb up to 27 times their dry weight in water. Today, peat moss is used mainly for a soil conditioner. A number of substances, such as phenols, waxes and resins, are extracted from peat moss for commercial use, and there is also the potential to produce ethyl or methyl alcohol.

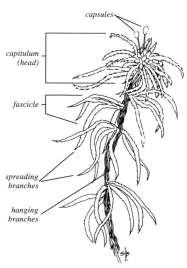

capsules

capitulum (head)

fascicle

spreading branches

hanging branches

Key to the Peat Mosses (*Sphagnum* spp.)

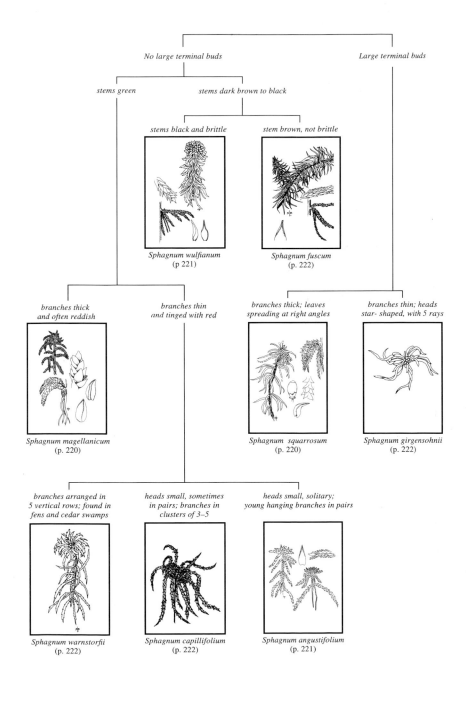

No large terminal buds   Large terminal buds

stems green   stems dark brown to black

stems black and brittle   stem brown, not brittle

Sphagnum wulfianum
(p 221)

Sphagnum fuscum
(p. 222)

branches thick
and often reddish

branches thin
and tinged with red

branches thick; leaves
spreading at right angles

branches thin; heads
star-shaped, with 5 rays

Sphagnum magellanicum
(p. 220)

Sphagnum squarrosum
(p. 220)

Sphagnum girgensohnii
(p. 222)

branches arranged in
5 vertical rows; found in
fens and cedar swamps

heads small, sometimes
in pairs; branches in
clusters of 3–5

heads small, solitary;
young hanging branches in pairs

Sphagnum warnstorfii
(p. 222)

Sphagnum capillifolium
(p. 222)

Sphagnum angustifolium
(p. 221)

S. magellanicum

S. magellanicum

COMMON PEAT MOSSES
Sphagnum spp., Section Sphagnum

GENERAL: Peat mosses with **sturdy thick branches and broad, hooded, overlapping branch leaves.**

Midway peat moss (*S. magellanicum*) is reddish (in the sun) to pink-tinged (in the shade). It grows in open bogs at lower and middle levels on hummocks, from Alaska to Greenland, south to California and Texas. Midway peat moss is a pioneer hummock former, invading poor fens. It commonly grows with **small red peat moss** (*S. capillifolium*, p. 222)

Central peat moss (*S. centrale*) forms light green carpets in wet hollows in cedar swamps and (less often) in treed fens, from Alaska to Newfoundland, south to Washington, Iowa and West Virginia.

Papillose peat moss (*S. papillosum*) forms golden-brown carpets in fens or around the edges of acidic lakes, from Ontario to Greenland, south to Minnesota and North Carolina in the east, and from Alaska to Oregon in the west. It is an indicator of fens that are influenced by groundwater, and it is usually succeeded by the red hummock formers, midway peat moss and small red peat moss.

S. magellanicum

SQUARROSE PEAT MOSSES
Sphagnum spp., Section Squarrosa

GENERAL: Light green or yellow-green peat mosses with large buds at stem tips and with branch leaves spreading at tips.

Shaggy peat moss (*S. squarrosum*) has branch leaves that are pressed against the stem at their bases, and that spread at right angles (squarrose) at their tips, giving the branches a bristly look. It grows in alkaline conditions in habitats such as swamps, stream edges and wet depressions in deciduous forests or boreal mixed-wood forests from Alaska to Greenland, south to California, Minnesota and North Carolina.

MANY-BRANCHED PEAT MOSSES
Sphagnum spp., Section Polyclada

GENERAL: Dark green, brown-tinged peat mosses with large, rounded heads (capitula), dark woody stems (break with a snap!) and branches in clusters (fascicles) of 6 or more.

Wulf's peat moss (*S. wulfianum*) is easily recognized by its large, shaggy, clover-like head and dark, brittle stems. It forms mounds over old logs and stumps or grows in loose mats in conifer (especially cedar) swamps, from British Columbia to Greenland, south to Alberta, Minnesota and Pennsylvania.

CUSPIDATE PEAT MOSSES
Sphagnum spp., Section Cuspidata

GENERAL: Peat mosses with branches in clusters (fascicles) of 5 or less, and growing in low, wet depressions.

Poor fen moss (*S. angustifolium*) has plants with small heads. It is typically green throughout, and it has young branches hanging in pairs. Poor fen moss forms loose carpets or lawns in shaded areas or hollows, or it grows scattered with other peat mosses on hummocks in bogs and poor fens, from Alaska to Newfoundland, south to Oregon, Wisconsin and West Virginia.

S. fuscum

S. capillifolium

S. girgensohnii

SHARP-LEAVED PEAT MOSSES
Sphagnum spp., Section Acutifolia

GENERAL: Green, brown or red peat mosses with branches in clusters (fascicles) of 5 or less, forming compact tufts.

Small red peat moss (*S. capillifolium*, also known as *S. nemoreum*) has reddish-green plants with round, often double, heads and narrow, oblong leaves with sharp-pointed, concave tips. It grows in open bogs (midheight on hummocks) and treed bogs (toward the bottom of hummocks), and it forms hummocks in fen mats that are in transition to bogs. Its range extends from Alaska to Greenland, south to Washington, Colorado, Arkansas and North Carolina. Small red peat moss commonly grows with **midway peat moss** (*S. magellanicum*, p. 220).

Wide-tongued peat moss (*S. russowii*) is easily distinguished from midway peat moss in the field. It differs from small red peat moss, because that species has swollen branch leaves, which give it a fatter appearance. Wide-tongued peat moss forms hummocks in bogs and fens, and occasionally in damp woods, from Alaska to Greenland, south to Washington, Colorado and North Carolina.

Common green peat moss (*S. girgensohnii*) is green, and it has a distinctive, 5-rayed, star-shaped head with a visible bud at the flattened tip. It forms carpets or mounds in swamps (especially cedar swamps), treed fens and wet depressions in hardwood or boreal mixed-wood forests, from Alaska to Greenland, south to Oregon, Minnesota and North Carolina. Common green peat moss can easily be confused with wide-tongued peat moss in the field, but that species often has reddish branch tips.

Warnstorf's peat moss (*S. warnstorfii*) has slender plants that are dark green when growing in the shade and bright red in the open. Its stems are usually reddish, and its leaves are arranged in 5 distinct rows, with spreading tips. It forms dense, low, mounds in rich fens and swamps (especially cedar swamps), from Alaska to Greenland, south to Oregon, Colorado and Florida.

Common brown peat moss (*S. fuscum*) forms dense brown or greenish-brown tufts. It has flat, narrow stem leaves with rounded tips, thread-like branches and brown stems. Common brown peat moss grows in open, acidic bogs on the dry tops of older, larger hummocks of other peat mosses, from Alaska to Greenland, south to Oregon, Colorado, Minnesota and Virginia. It occasionally forms large, isolated mounds in rich fens.

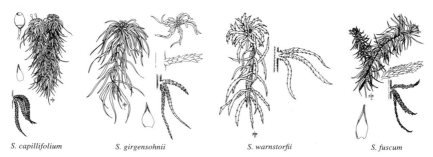

S. capillifolium   *S. girgensohnii*   *S. warnstorfii*   *S. fuscum*

APPENDIX
Scientific names of animal species mentioned in this guide

Birds

American black duck .... *Anas rubripes*
American bittern .... *Botaurus lentiginosus*
American coot .... *Fulica americana*
American crow .... *Corvus brachyrhynchos*
American goldfinch .... *Carduelis tristis*
American robin .... *Turdus migratorius*
American tree sparrow .... *Spizella arborea*
American woodcock .... *Scolopax minor*
bobolink .... *Dolichonyx oryzivorus*
bufflehead .... *Bucephala albeola*
Canada goose .... *Branta canadensis*
canvasback .... *Aythya valisineria*
cedar waxwing .... *Bombycilla cedrorum*
chickadees .... *Parus atricapillus* and *P. hudsonicus*
common goldeneye .... *Bucephala clangula*
common redpoll .... *Carduelis flammea*
common snipe .... *Gallinago gallinago*
eastern bluebird .... *Sialia sialis*
European starling .... *Sturnus vulgaris*
evening grosbeak .... *Coccothraustes vespertinus*
grey catbird .... *Dumetella carolinensis*
mallard .... *Anas platyrhynchos*
marsh wren .... *Cistothorus palustris*
northern oriole .... *Icterus galbula*
northern pintail .... *Anas acuta*
pied-billed grebe .... *Podilymbus podiceps*
pine siskin .... *Carduelis pinus*
purple finch .... *Carpodacus purpureus*
red-winged blackbird .... *Agelaius phoeniceus*
redhead .... *Aythya americana*
ring-necked duck .... *Aythya collaris*
ruby-throated hummingbird .... *Archilochus colubris*
ruffed grouse .... *Bonasa umbellus*
scaup .... *Aythya marila* and *A. affinis*
sora .... *Porzana carolina*
swamp sparrow .... *Melospiza georgiana*
tundra swan .... *Cygnus columbianus*
teal .... *Anas crecca* and *A.discors*
tree swallow .... *Tachycineta bicolor*
wood thrush .... *Hylocichla mustelina*
wood duck .... *Aix sponsa*

Mammals

beaver	*Castor canadensis*
black bear	*Ursus americanus*
chipmunks	*Tamias striatus* and *Eutamias minimus*
cottontail rabbit	*Sylvilagus floridanus*
meadow vole	*Microtus pennsylvanicus*
moose	*Alces alces*
muskrat	*Ondatra zibethicus*
porcupine	*Erethizon dorsatum*
raccoon	*Procyon lotor*
red fox	*Vulpes vulpes*
red squirrel	*Tamiasciurus hudsonicus*
river otter	*Lontra canadensis*
rock vole	*Microtus chrotorrhinus*
snowshoe hare	*Lepus americanus*
striped skunk	*Mephitis mephitis*
white-tailed deer	*Odocoileus virginianus*

Fish

muskellunge	*Esox masquinongy*
northern pike	*Esox lucius*

Invertebrates

ambush bugs	Order Hemiptera, Family Phymatidae
ants	Order Hymenoptera, Family Formicidae
Baltimore butterfly	*Euphydryas phaeton*
bees	Order Hymenoptera
beetles	Order Coleoptera
bumblebees	*Bombus* spp.
cattail moth	*Bellura obliqua*
flies	Order Diptera
goldenrod beetle	Order Coleoptera
hawk moths	Order Lepidoptera, Family Sphingidae
iris borer	*Macronoctua onusta*
leafhoppers	Order Homoptera, Family Cicadellidae
midges	Order Diptera, Family Chironomidae
milkweed bug	*Lygaeus kalmii*
milkweed beetle	*Tetraopes* sp.
monarch butterfly	*Danaus plexippus*
mosquitos	Order Diptera, Family Culicidae
moths	Order Lepidoptera
pitcher plant mosquito	*Wyeomyia smithii*
weevils	Order Coleoptera, Family Curculionidae

GLOSSARY

achene: a small, dry, 1-seeded, nut-like fruit.

acrocarpus: producing archegonia and sporophytes at the stem tips.

adventitious: not growing from the usual place.

annual: completing the life cycle in 1 year.

anther: a pollen-bearing portion of a stamen.

antheridium [antheridia]: the male sex organ of a non-flowering plant.

apex [apices]: a tip.

apical: at the tip.

appressed: lying close to or flat against a surface.

aquatic: adapted to living partially to wholly submerged in water or in waterlogged soil.

archegonium [archegonia]: the female sex organ of a non-flowering plant.

auricle: an ear-shaped lobe.

awn: a slender bristle, often at the tip of a leaf.

axil: the angle between a leaf and a stem.

axillary: in an axil.

axis: the central, longitudinal line, along which parts of the plant are arranged.

bi-: twice- or 2-.

bilaterally symmetrical: vertically divisible into equal halves (compare radially symmetrical).

bisexual: with both sexes on the same individual.

biennial: living for 2 years, usually flowering and producing fruit in the 2nd year.

biodiversity: the full range of life in all its forms—including genes, species and ecosystems—and the ecological processes that link them.

blade: the expanded portion of a leaf.

bog: a waterlogged, peat-forming area, dominated by *Sphagnum* mosses, and poor in cations and nutrients because all water is derived from precipitation.

bract: a specialized, reduced leaf associated with a flower or flower cluster.

branchlet: a small branch.

bulb: a short, vertical, underground stem thickened by overlapping, fleshy leaves or leaf-bases.

calcareous: calcium-rich; rich in lime.

calciphile: a plant that prefers calcareous habitats.

calyptra [calyptrae]: a thin covering or hood over a developing moss capsule, usually shed at maturity.

calyx: the sepals collectively.

cambium: a layer of tissue, usually 1 cell thick, that produces the conducting tubes (xylem and phloem).

capsule: in seed plants, a dry fruit composed of more than 1 carpel and splitting open at maturity; in non-flowering plants, a spore-containing sac.

carpel: a basic, female reproductive structure, consisting of an ovary, a style and a stigma.

cathartic: laxative, emptying the bowels.

catkin: a linear cluster of small flowers with a single bract and no petals, usually male or female.

chlorophyll: the green, photosynthetic pigment of plants.

ciliate: fringed with hairs or hair-like structures.

circumpolar: occurring around the world.

clambering: trailing over the ground.

clasping: surrounding the stem, usually referring to leaf bases.

cleft: deeply lobed; cut about halfway to the midrib or base, or a little deeper.

compound: divided into smaller parts; leaves divided into leaflets or flower clusters divided into smaller groups.

convex: rounded on the surface.

corolla: the petals of a flower, collectively.

corm: a short, vertical, thickened underground stem without thickened leaves.

cortex: the outer layer of a thallus or stem.

crisped: irregularly curled or rippled along the edges, like a piece of fried bacon.

culm: a plant stem.

cultivar: a cultivated variety.

cyme: a flower cluster in which the flowers at the branch tips open first.

deciduous: falling after completion of its normal function, often at the approach of a dormant season.

dehiscence: the act or method of opening, as in fruit, anthers or spore capsules.

dehiscent: opening by a definite pore(s) or along a regular line(s) to discharge seeds or spores.

diaphoretic: stimulating perspiration.

dichotomous: branching in 2 equal parts, like a "Y."

dioecious: producing male and female parts on separate individuals.

dimorphic: occurring in 2 forms, as in ferns with sterile and fertile leaves.

disc floret: a tubular flower of a plant in the aster family.

disjunct: disconnected; in biogeography, referring to a population that is separated (distant) from all other populations of that species.

dorsal: on the outer side of a structure (the surface away from the axis).

drupe: a fleshy or pulpy, 1-seeded fruit, in which the seed has a stony covering.

elongate: considerably longer than wide.

emergent: coming out from; only partly submersed in water.

Parts of a Typical Flower

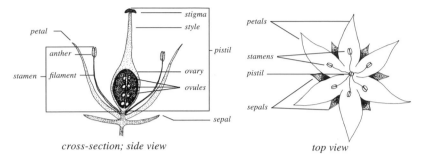

cross-section; side view

top view

Flower Arrangement

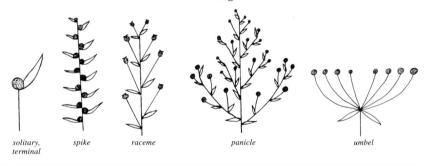

solitary, terminal

spike

raceme

panicle

umbel

---

endemic: growing only in a particular region.

entire: without teeth or lobes; with a continuous edge.

epiphyte: a plant growing on another plant.

ericaceous: belonging to the family Ericaceae and usually having thick leathery leaves.

fascicle: a small bundle or cluster.

fen: a peat-forming area receiving nutrients from ground water and precipitation, and therefore richer in cations and nutrients than a bog.

floodplain: a plain built up by stream deposits through repeated flooding.

floret: a tiny flower, usually part of a cluster.

frond: a fern leaf.

gametophyte: a plant that produces sexual reproductive structures; in bryophytes, the green, leafy or thalloid plants.

glandular: with glands.

glaucous: with a whitish, greyish or bluish coating resembling the waxy bloom on a plum.

glume: 1 of 2 empty bracts at the base of a grass spikelet.

graminoid: a grass-like plant; a grass, sedge or rush.

head: a short, condensed cluster.

herb: a plant without woody, above-ground parts, the stems dying back to the ground each year.

herbaceous: herb-like.

hummock: a raised mound, often in a bog or fen.

hyaline: translucent or transparent.

hybridization: the process of creating a hybrid by cross-breeding different species.

incurved: curved upwards and inward.

indusium: an outgrowth covering and protecting a spore cluster on a fern leaf.

inflorescence: a flower cluster. The arrangement of the flowers on the axis.

inflorescence unit: In sedges, the flower cluster can be divided into "terminal" or "lateral" inflorescence units. These subdivisions are made up of spikes, which are further divided into spikelets (the actual perigynia and scale).

internode: a portion of a stem between adjacent nodes.

involute: with edges rolled upwards to the top of a leaf.

irregular flower: a flower with petals (or less often sepals) dissimilar in form or orientation.

julaceous: smoothly cylindrical due to closely and evenly overlapping leaves.

keel: a sharp, lengthwise ridge like the keel of a boat, as when a leaf is folded along its midrib.

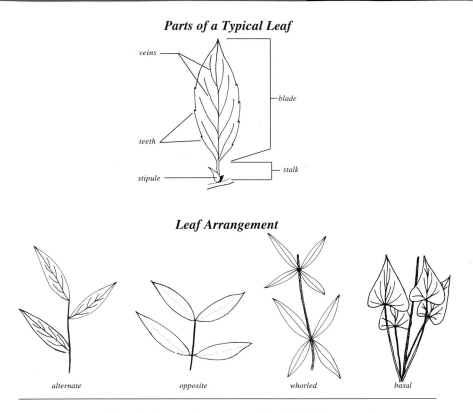

Parts of a Typical Leaf

veins

blade

teeth

stipule

stalk

Leaf Arrangement

alternate opposite whorled basal

lawn: a continuous, relatively level expanse of mosses.

lemma: the lower of the 2 bracts of a single grass flower.

lenticel: a slightly raised pore on root or stem bark.

ligule: the flat, usually membranous, collar-like projection at the upper edge of a grass leaf sheath; the strap-shaped part of a ray flower of the aster family.

linear: leaf several times larger than wide with parallel edges.

lip: a projection or expansion of something, such as the lower petal of an orchid or violet flower.

lobe: a rounded or strap-shaped division of the leaf.

midrib: the central rib of a leaf.

midvein: the central vein of a leaf.

monoecious: with separate male and female flowers on the same plant.

nectary: a gland that secretes nectar.

node: a joint; the place where a leaf or branch is attached to a stem.

oblique: an angle of attachment between 0° and 90° to the stem.

ocrea (ochrea): a membranous tissue like the sheath found at the nodes of the stem in the smartweed family (Polygonaceae).

operculum: the lid of a moss capsule.

opposite: situated across from each other, not alternate or whorled; situated directly in front of an organ of another kind.

ovary: the structure that encloses the young, undeveloped seeds (ovules).

palea: the upper of the 2 bracts of a single grass flower.

palmate: divided into lobes or leaflets diverging from a common point, like fingers on a hand.

papilla [papillae]: a minute, wart-like projection.

pappus: the cluster of bristles of long hairs at the top of an achene of the aster family.

parasite: an organism that gets its nourishment chiefly from a living organism to which it is attached.

peat: organic matter accumulated as a result of incomplete decomposition, usually in water-soaked conditions.

pendent: hanging down.

perennial: living for more than 2 years.

perianth: the sepals and petals of a flower, collectively; in liverworts, a tube of 2–3 fused leaves surrounding a developing sporophyte.

Leaf Tips

acuminate *acute* *cuspidate*

---

perichaetial leaves or **perichaetium** [perichaetia]: the special leaves or bracts surrounding the female reproductive organ of a moss.

perigonial leaves or **perigonium** [perigonia]: the special leaves or bracts surrounding the male reproductive organ of a moss.

perigynium [perigynia]: an inflated sac enclosing the achene of a sedge flower.

petal: 1 of the inside ring of modified flower leaves, often white or brightly coloured.

petiole: a leaf stalk.

photosynthesis: the process by which green plants produce their food (carbohydrates) from water, carbon dioxide and minerals, using the sun's energy.

pinna: the primary division of a pinnately compound leaf.

pinnate: with leaflets arranged on 2 sides of a central axis, like the barbs of a feather.

pinnatifid: with lobes arranged on 2 sides of a central axis.

pinnule: a division of a pinna.

pistil: the female organ of a flower, usually consisting of an ovary, style and stigma.

plumose: feathery.

rachis: the main axis of a compound leaf.

radially symmetrical: developing uniformly on all sides, like spokes on a wheel.

rank: a row, as a vertical row of leaves on a stem.

ray floret: a flower of the aster family with a strap-like (ligulate) corolla, often produced at the outer edge of the flowerhead.

receptacle: the enlarged end of a stem to which the flower parts (or, in the aster family, the florets) are attached.

recurved: bent backward.

regular flower: a flower with petals and sepals similar in size, shape, and orientation.

remote: widely spaced.

rhizoid: a slender, simple or branched, root-like growth on a bryophyte, that absorbs nutrients and anchors the plant.

rhizome: an underground stem, usually elongated.

riparian: adjacent to a river or stream, including shores and floodplains.

rosette: a roughly circular cluster radiating from a central point.

rugose: wrinkled like crumpled paper.

runner: a slender stolon.

samara: a dry, usually winged fruit.

saprophyte: an organism living on dead organic matter, neither parasitic nor making its own food.

scale: a small, thin, flattened structure.

sedge: a grass-like plant with solid, frequently triangular stems, closed leaf sheaths and achenes.

serrulate: with fine, sharp teeth.

sessile: without a stalk.

seta [setae]: the stalk of a moss or liverwort capsule.

sheath: an organ that partly or completely surrounds another organ, as the sheath of a grass leaf surrounds the stem.

simple: not divided or subdivided.

sorus [sori]: a cluster of small spore cases (sporangia) on the underside of a fern leaf.

spicule: a small, slender, sharp-pointed body.

spike: An elongate inflorescence with stalkless flowers. In grasses, spikes can be made up of several spikelets. In sedges, the inflorescence unit may be divided into male and female spikes.

spikelet: The smallest unit of an inflorescence. In a grass, a spikelet is made up of 1 to several flowers. In sedges, the spikelet is a perigynia with its scale.

spinulose: covered by tiny spines.

Leaf Edges

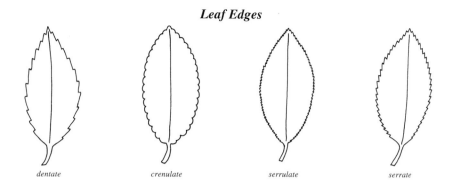

dentate   *crenulate*   *serrulate*   *serrate*

---

sporangium [sporangia]: a spore case.

spore: a small, haploid, 1–few-celled reproductive body produced in a capsule (mosses and liverworts) or sporangium (ferns) and capable of giving rise to a new plant.

sporophyll: a spore-bearing leaf.

sporophyte: the diploid, spore-bearing part or phase of a plant, in bryophytes, the capsule, with its stalk and foot; in ferns the leafy plant.

spur: a hollow appendage on a petal or sepal.

squarrose: spreading at right angles.

stamen: the pollen-bearing organ of a flower.

staminodium: a sterile stamen, often with a modified structure.

stigma: the tip of the female organ of a flowering plant, where the pollen lands.

stipule: an appendage at the base of a leaf stalk, usually leaf-like.

stolon: a stem or runner spreading horizontally on the ground, usually rooting at nodes or tips.

stoma, stomate [stomata]: a tiny pore in a plant surface, bounded by a pair of guard cells which can close the opening by changing shape.

style: the stalk, or middle part, of the female organ of a flowering plant, connecting the stigma and ovary.

submergent: under the water.

substrate: the surface on which something grows.

succession: the process of change in plant community composition and structure over time.

symbiosis: a mutually advantageous partnership between 2 dissimilar organisms.

taproot: a primary descending root.

taxonomy: the theory and practice of describing organisms and ordering them in a system of classification.

tepal: a sepal or a petal, when these structures are essentially identical and not easily distinguished.

terete: circular in cross-section.

terminal: at the end, or tip of.

terrestrial: living on or growing in the earth.

thallose: with a thallus.

thallus: a plant body that is not differentiated into a stem and leaves as in some hepatics.

tubercle: a small swelling or projection on an organ.

tubular: in leaves, when margins are inrolled and overlapping to form a tube.

turion: a scaly, bulb-like growth from a bud on a rhizome.

tussock: a compact tuft of grasses or sedges.

unarmed: without spines, prickles or thorns.

undulate: with a wavy edge or surface, the waves oriented across the organ, rather than lengthwise.

unisexual: with organs of only 1 sex; male or female.

valve: 1 of the pieces into which a pod or capsule splits.

vein: a strand of conducting tubes (a vascular bundle), often visible externally in leaves.

ventral: on the upper or inner surface of a leaf; opposite to dorsal.

voucher specimen: a specimen collected for verification of identification.

wetland: a permanently or seasonally waterlogged area, such as a marsh, swamp, bog or fen.

whorl: a ring of 3 or more similar structures (e.g. leaves) arranged around an axis at the same level.

wing: a thin, flattened extension or projection from the side or tip of an organ; 1 of the 2 side petals of a flower in the pea family.

winter bud: a specialized, overwintering, vegetative bud that resumes growth in the spring, usually produced in late summer.

REFERENCES

Argus, G.W. 1964. *Preliminary reports on the flora of Wisconsin. No. 51. Salicaceae. The genus* Salix—*the Willows.* Wisconsin Academy of Science, Arts and Letters, Madison.

Argus, G.W., K. M. Pryer, D.J. White and C.J. Keddy. 1987. *Atlas of the Rare Vascular Plants of Ontario.* 4 vols. National Museum of Natural Sciences, Ottawa.

Assiniwi, B. 1972. *Survival in the Bush.* Copp Clark Publishers, Toronto.

Baldwin, K.A., and R.A. Sims. 1989. *Common Forest Plants in Northwestern Ontario.* Ontario Ministry of Natural Resources, Toronto.

Baldwin, W.K. 1958. *Plants of the Clay Belt of Northern Ontario and Quebec.* Bulletin No. 156. National Museums of Canada, Ottawa.

Banfield, A.W.F. 1974. *The Mammals of Canada.* University of Toronto Press. Toronto.

Berglund, B. 1974. *Wilderness Survival.* Pagurian Press, Toronto.

Blackmore, S., and E. Tootill. 1988. *The Penguin Dictionary of Botany.* Penguin Books, London, England.

Boericke, W. 1927. *Materia Medica.* 9th ed. Boericke and Tafel, Santa Rosa, California.

Boivin, B. 1992. *Les Cyperaceae de l'est du Canada.* Provancheria, Memoire de L'Herbier Lois-Marie, no. 25. Université Laval, Quebec.

Borror, D.J., and R.E. White. 1970. *A Field Guide to the Insects of America North of Mexico.* Houghton Mifflin Company. Boston.

Britton, N., and A. Brown. [1913] 1970. *An Illustrated Flora of the Northern United States and Canada.* 3 vols. Dover Publications, New York.

Brown, L. 1979. *Grass Identification Guide.* Houghton Mifflin Company, Boston.

Cadman, M.D., P.F.J. Eagles and F.M. Helleiner. 1987. *Atlas of the Breeding Birds of Ontario.* University of Waterloo Press, Waterloo, Ontario.

Cody, C.J., and D.M. Britton. 1989. *Ferns and Fern Allies of Canada.* Agriculture Canada, Ottawa.

Coffin, B., and L. Pfannmuller, eds. 1988. *Minnesota's Endangered Flora and Fauna.* University of Minnesota Press, Minneapolis.

Coombes, A.J. 1989. *Dictionary of Plant Names.* Timber Press, Portland, Oregon.

Crow, G.E., and C.B. Hellquist. 1985. *Aquatic Vascular Plants of New England.* 8 parts. New Hampshire Agricultural Experiment Station, University of New Hampshire, Ourham.

Crowe, J.M. 1992. *The Liverworts of Southwest Thunder Bay District: A Concise Hepatic Flora.* Claude E. Garton Herbarium, Lakehead University, Thunder Bay, Ontario.

Crum, H.A., and L. E. Anderson. 1981. *Mosses of Eastern North America.* 2 vols. Columbia University Press, New York.

Crum, H.A. 1983. *Mosses of The Great Lakes Forest.* University of Michigan Press, Ann Arbor.

———. 1991. *Liverworts and Hornworts of Southern Michigan.* University of Michigan Press, Ann Arbor.

———. 1992. *A Focus on Peatlands and Peat Mosses.* University of Michigan Press, Ann Arbor.

Damman, A.W.H. 1963. *Key to the* Carex *Species of Newfoundland by Vegetative Characteristics.* Department of Forestry, Canadian Forest Service, Ottawa.

Davis, P.H., and J. Cullen. 1989. *The Identification of Flowering Plant Families.* 3rd ed. Cambridge University Press, New York.

Dobson, I., and P.M. Catling. (no date). *Pondweeds of the Ottawa District.* Biosystematics Research Institute, Agriculture Canada, Ottawa.

Dore, D.G., and J. McNeil. 1980. *Grasses of Ontario.* Agriculture Canada, Ottawa.

Erichsen-Brown, C. 1979. *Medicinal and Other Uses of North American Plants: A Historical Survey with Special Reference to the Eastern Indian Tribes.* Dover Publications, New York.

Fassett, N.C. 1957. *A Manual of Aquatic Plants.* University of Wisconsin Press, Madison.

Gaevskaia, N.S. 1969. *The Role of Higher Aquatic Plants In The Nutrtion of the Animals of Freshwaters.* National Library for Science and Technology, Moscow, U.S.S.R.

Garton, C. 1994. *Checklist of the Plants of Thunder Bay District.* Rev. by J. Crowe. Thunder Bay Field Naturalists, Thunder Bay, Ontario.

Gillett, J.M., and N.K.B. Robson. 1981. *The St. John's-worts of Canada (Guttiferae).* Publications in Botany, no. 11. National Museum of Natural Sciences, Ottawa.

Gillett, J.M., and D.J. White. 1978. *Checklist of Vascular Plants of the Ottawa-Hull Region, Canada.* National Museum of Natural Sciences, Ottawa.

Gleason, H.A., and A. Cronquist. 1991. *Manual of Vascular Plants of United States and Adjacent Canada.* New York Botanical Gardens, New York.

Gledhill, D. 1990. *The Names of Plants.* Cambridge University Press, New York.

Godfrey, W.E. 1986. *The Birds of Canada.* Rev. ed. National Museum of Natural Sciences, Ottawa.

Hosie, R.C. 1979. *Native Trees of Canada.* Fitzhenry and Whiteside, Don Mills, Ontario.

Hotchkiss, N. 1972. *Common Marsh, Underwater and Floating-leaved Plants of the United States and Canada.* Dover Publications, New York.

Hutchens, A.R. 1991. *Indian Herbology of North America: The Definitive Guide to Native Medicinal Plants and Their Uses.* Shambhala Publications, Boston.

———. 1992. *A Handbook of Native American Herbs.* Shambhala Publications, Boston.

Ireland, R.R., C.D. Bird, G.R. Brassard, W.B. Schofield and D.H. Vitt. 1980. *Checklist of the Mosses of Canada.* Publications in Botany, no. 8. National Museums of Canada, Ottawa.

Ireland, R.R., and B.T. Gilda. 1987. *Illustrated Guide To Some Hornworts, Liverworts and Mosses.* Syllogeus No. 62. National Museums of Canada, Ottawa.

Ireland, R.R., and L.M. Ley. 1992. *Atlas of Ontario Mosses.* Syllogeus No. 70. Canadian Museum of Nature, Ottawa.

James, R.D. 1991. *Annotated Checklist of the Birds of Ontario.* 2nd ed. Miscellaneous Publications. Royal Ontario Museum, Toronto.

Jeglum, J.K., A.N. Boissonneau and V.F. Haavisto. 1974. *Toward a Wetland Classification for Ontario.* Information Report O-X-215. Canadian Fororest Service, Sault Ste. Marie, Ontario.

Johnston, B. 1978. *Ojibway Language Lexicon.* Department of Ethnology, Royal Ontario Museum, Toronto.

Kowalchik, C., and W.H. Hylton. *Rodales Illustrated Encyclopedia of Herbs.* Rodale Press, Emmaus, Pennsylvania.

Lakela, O. 1965. *A Flora of Northeastern Minnesota.* University of Minnesota Press, Minneapolis.

Larson, G.E. 1993. *Aquatic and Wetland Vascular Plants of the Northern Great Plains.* General Technical Report RM-238. Rocky Mountain Forest and Range Experiment Station, Forest Service, United States Department of Agriculture, Fort Collins, Colorado.

Lellinger, D.B. 1985. *Ferns and Fern Allies of Canada.* Smithsonian Institute Press, Washington, D.C.

Mackinnon, A., J. Pojar and R. Coupé. 1992. *Plants of Northern British Columbia.* Lone Pine Publishing, Vancouver.

Magee, D.M. 1981. *Freshwater Wetlands: A Guide to Common Indicator Plants of the Northeast.* University of Massachusetts Press, Amherst.

Marie-Victorin, F. 1964. *Flore Laurentienne.* C.P. 6128. Les Presses de L'Université de Montreal, Montreal.

Martin, A.C., H.S. Zim and A.L. Nelson. 1951. *American Wildlife and Plants: A Guide to Wildlife Food Habits.* Dover Publications, New York.

Medsger, O.P. 1966. *Edible Wild Plants.* Collier Macmillan Publications, New York.

Flora of North America Editorial Committee. 1993. *Flora of North America.* Vol. 2. Oxford University Press, New York.

Morton, J.K., and J.M. Venn. 1984. *The Flora of Manitoulin Island.* 2nd ed. Biology Series, no. 28. University of Waterloo Press, Waterloo, Ontario.

———. 1990. *A Checklist of the Flora of Ontario Vascular Plants.* Biology Series, no. 34. University of Waterloo Press, Waterloo, Ontario.

Mulligan, G.A., and D.B. Munroe. 1990. *Poisonous Plants of Canada.* Agriculture Canada, Ottawa.

National Wetlands Working Group. 1988. *Wetlands of Canada.* Ecological Land Classification Series, no. 24. Sustainable Development Branch, Environment Canada, Ottawa.

Newcomb, L. 1977. *Newcomb's Wildflower Guide.* Little, Brown and Company, Toronto.

Newmaster, S.G., A. Lehela, M.J. Oldham, P.W.C. Uhlig and S. McMurray. 1996. *Ontario Plant List.* Ontario Forest Research Institute, Sault Ste. Marie. Forest Research Paper No. 123.

Oldham, M.J. 1994. *Rare Vascular Plants.* Natural Heritage Information Centre, Peterborough, Ontario.

Ontario Minstry of Natural Resourses. 1993. *Ontario Wetland Evaluation System: Northern Manual.* Ontario Ministry of Natural Resources, Toronto.

Panos, M.B., and J. Heimlich. 1980. *Homeopathic Medicine at Home.* J.P. Tarcher, New York.

Payne, N.F. 1992. *Techniques for Wildlife Habitat Management of Wetlands.* McGraw-Hill, New York.

Peterson, L.A. 1977. *A Field Guide to Edible Wild Plants: Eastern and Central North America.* Houghton Mifflin Company, Boston.

Peterson, R.L. 1966. *The Mammals of Eastern Canada.* Oxford University Press, Toronto.

Peterson, R.T. 1968. *A Field Guide to Wildflowers.* Houghton Mifflin Company, Boston.

———. 1980. *A Field Guide to the Birds of Eastern and Central North America.* Houghton Mifflin Company, Boston.

Raven, P.H., R.F. Evert and S.E. Eichhorn. 1986. *Biology of Plants.* Worth Publishers, New York.

Reznicek, A.A. 1990. Evolution in Sedges (*Carex*, Cyperaceae). *Canadian Journal of Botany* 68:1409–1432.

———. 1994. The disjunct coastal plain flora in the Great Lakes Region. *Biological Conservation* 68:203–215.

Riley, J.L. 1989. *Distribution and Status of the Vascular Plants of Central Region, Ontario.* Ontario Ministry of Natural Resources, Toronto.

Santillo, H. 1987. *Natural Healing With Herbs.* Hohm Press, Prescott Valley, Arizona.

Scoggan, H.J. 1978. *The Flora of Canada.* 4 vols. Publications in Botany, no. 7. National Museum of Natural Sciences, Ottawa.

Scott, W.B., and E.J. Crossman. 1973. *Freshwater Fishes of Canada.* Bulletin 184. Fisheries Research Board of Canada, Ottawa.

Sculthrope, C.D. 1967. *The Biology of Aquatic Vascular Plants.* Edward Arnold, London, England.

Semple, J.C., and S.B. Heard. 1987. *The Asters of Ontario:* Aster *L. and* Virgulus *Raf. (Compositae:Asteraceae).* Department of Biology, University of Waterloo Press, Waterloo, Ontario.

Semple, J.C., and G.S. Ringius. 1983. *The Goldenrods of Ontario:* Solidago *L. and* Euthamia *Nutt.* Department of Biology, University of Waterloo Press, Waterloo, Ontario.

Skelton, E.G., and E.W. Skelton. 1991. *Haliburton Flora.* Life Sciences Miscellaneous Publications. Royal Ontario Museum, Toronto.

Smith, J.P. 1977. *Vascular Plant Families.* Mad River Press, Eureka, California.

Soper, J.H., and M.L. Heimburger. 1982. *Shrubs of Ontario.* Royal Ontario Museum, Toronto.

Soper, J.H., C.E. Garton and D. Given. 1989. *Flora of the North Shore of Lake Superior.* Syllogeus No. 63. National Museum of Natural Sciences, Ottawa.

Stearn, W.T. 1983. *Botanical Latin.* Fitzhenery and Whiteside, Toronto.

Stokes, D., and L. Stokes. 1985. *A Guide to Enjoying Wildflowers.* Little, Brown and Company, Toronto.

Szczawinski, A.F., and N.J. Turner. 1979. *Edible Wild Plants of Canada Series,* Nos. 3 & 4. National Museum of Natural Sciences, Ottawa.

Traupman, J.C. 1988. *Latin and English Dictionary.* Bantam Books, Toronto.

Vander Kloet, S.P. 1988. *The Genus* Vaccinium *in North America.* Publication 1828. Agriculture Canada, Ottawa.

Voss, E.G. 1972. *Michigan Flora.* Vol. 1. University of Michigan Herbarium. University of Michigan Press, Ann Arbor.

———. 1985. *Michigan Flora.* Vol. 2. University of Michigan Herbarium. University of Michigan Press, Ann Arbor.

Walshe, S. 1980. *Plants of Quetico and the Ontario Shield.* University of Toronto Press, Toronto.

Walters, D.R., and D.J. Keil. 1977. *Vascular Plant Taxonomy.* 3rd ed. Kendal/Hunt Publications, Dubuque, Iowa.

Ward, H.B., and G.C. Whipple. 1918. *Fresh-water Biology.* John Wiley and Sons, London, England.

Whiting, R.E., and P.M. Catling. 1986. *Orchids of Ontario.* The CanaColl Foundation, Agriculture Canada, Ottawa.

PHOTOGRAPH AND ILLUSTRATION CREDITS

Photographs

All photographs are used with the generous permission of their copyright holders. The numbers refer to pages; the letters indicate the photograph's relative position on the page (ordered from left to right, top to bottom).

Brenda Chambers: 17c, 19, 21a, 23a, 25c, 31b, 39b, 56b, 59a, 73c, 88b, 207, 216b.

D. R. Gunn: 27c, 56c, 63d, 63e, 64a, 64b, 65c, 65d, 65e, 75a, 75c, 94a, 100c, 151a, 151b, 162a.

Allan Harris: 9b, 9c, 9d, 9e, 9f, 10a, 10d, 10e, 10f, 17a, 18a, 18b, 28b, 39a, 58b, 62b, 62c, 73b, 78, 80c, 84b, 87a, 100a, 102c, 104, 112c, 112d, 113a, 113b, 113c, 118a, 118b, 119a, 120a, 121a, 122a, 123a, 123b, 125c, 126a, 126b, 127b, 128b, 129b, 130, 131a, 132a, 133, 134b, 136a, 137, 138b, 139a, 139b, 139c, 140a, 140b, 142a, 142b, 142c, 143, 144b, 145a, 146a, 146b, 147a, 148b, 150a, 157b, 158b, 159a, 159b, 160a, 160b, 160c, 163a, 163c, 167c, 171a, 171b, 174b, 175b, 178a, 178b, 179a, 179b, 180a, 181c, 182a, 183a, 183b, 184b, 184c, 187a, 187b, 188a, 199a.

Derek Johnson: 38b, 38c, 61a, 68, 80b, 81b, 96, 101b, 135, 169b, 177, 185a, 188c, 220a.

Linda Kershaw: 14b, 18c, 21b, 25b, 30a, 36b, 39c, 42b, 45a, 47b, 54a, 54c, 58a, 67b, 70a, 74b, 76a, 77c, 83b, 85c, 89b, 91b, 108a, 125a, 128a, 142d, 149b, 160d, 164b, 181b, 183c, 187c, 193a, 193b, 196b, 199b, 201b, 208a, 209b, 212b, 217b, 217c.

Harold Kish: 20b, 29a, 32b, 33b, 34b, 35b, 38a, 44, 46a, 46b, 48, 56a, 57a, 57c, 58c, 59b, 60a, 60b, 61b, 62a, 63a, 63b, 63c, 65a, 65b, 69a, 69b, 70b, 71b, 72b, 73a, 75b, 76b, 77b, 89a, 89c, 90a, 90b, 90c, 91a, 91c, 93a, 95b, 97b, 98b, 100b, 101a, 110a, 111b, 122b, 124a, 124b, 129a, 136a, 144a, 145c, 152, 161a, 161b, 163b, 166a, 166b, 167a, 167b, 182b, 186b, 190, 192b, 194b, 194c, 198b, 201d.

Steve Newmaster: 13, 14a, 15b, 16b, 24a, 25a, 27a, 28a, 31a, 32a, 33a, 34a, 35a, 36a, 43a, 45b, 47a, 55b, 57b, 71a, 72c, 74a, 77a, 79b, 80a, 83a, 84a, 86, 87b, 94b, 95a, 97a, 98a, 98c, 99a, 102a, 103, 108b, 108c, 111a, 114a, 127a, 134a, 138a, 141a, 148a, 148c, 150b, 156a, 156b, 164a, 165, 173b, 175a, 181a, 186a, 189a, 193c, 194a, 197a, 198a, 200a, 201a, 201c, 202, 208b, 209a, 210, 212a, 213a, 213b, 214, 215a, 215b, 216a, 217a, 220b, 220c, 221b, 222a, 222b, 222c.

Mike Oldham: 17b, 29c, 62d, 64c, 79c, 81a, 85a, 85b, 93b, 115c, 119b, 136b, 147b, 162b, 170a, 182c, 188b, 197b.

Ontario Ministry of Natural Resources: 20a, 24b, 82b, 115a, 125b, 129c, 145b, 157a, 172, 173a, 174a, 176a, 178c, 189b, 195c.

Other Photographers: **Wasyl Bakowsky**: 26b, 92a, 107b, 141b; **Frank Boas**: 30b, 37a, 37b, 82a, 85d, 123c; **Mike Bryan**: 40a, 40b, 42a, 54b, 67c, 110b, 168; **Sue Bryan**: 66, 109a; **Adolf Ceska**: 149a; **Bill Crins**: 22, 200b; **Cathy Paroschy Harris**: 115b; **Karen Legasy**: 15a, 16a, 26a, 29b, 41b, 43b, 72a, 88a, 99b; **Jennifer Line**: 10c, 55a, 107a, 121b, 192a, 221a; **Robert Norton**: 114c; **Jim Pojar**: 41a; **Gerry Racey**: 9a; **Anna Roberts**: 109b, 114b, 120b, 120c; **Sleeping Giant Provincial Park**: 60c; **Richard Sims**: 10b; **Trygve Steen**: 67a, 112a, 112b, 170b, 176b, 180b, 185b, 205; **Donald Sutherland**: 27b, 131b, 136b, 182d, 195a, 195b; **Thunder Bay Field Naturalists**: 102b, 184a, 196a; **Emma Thurley**: 71c; **Shan Walshe**: 23b, 79a, 92b, 132b, 158a, 169a.

Illustrations

All illustrations are used with the generous permission of their copyright holders. Numbers refer to pages.

Annalee McColm: 13, 14, 15, 16, 18, 19, 25, 31, 32, 34, 36, 38, 40, 42, 43, 49, 57, 58, 67, 71, 76, 89, 100, 106, 110, 116, 121, 123, 125, 136, 191, 193, 198, 200.

From **Britton and Brown**. [1913] 1970. *An Illustrated Flora of the Northern United States and Canada. Vol. 1: Ferns to Buckwheat.* Reprint of 2nd ed. Dover Publications, New York: 19, 21, 23, 66, 106, 108, 109, 110, 111, 112, 113, 123, 124, 126, 130, 136, 137, 141, 142, 143, 147, 149, 151, 159, 160, 162, 169, 172, 197.

From **Britton and Brown**. [1913] 1970. *An Illustrated Flora of the Northern United States and Canada. Vol. 2: Amaranth to Polypremum.* Reprint of 2nd ed. Dover Publications, New York: 30, 35, 68, 68, 74, 78, 80, 82, 84, 85, 154, 155, 158, 165, 177, 180, 184, 185, 188.

From **Britton and Brown**. [1913] 1970. *An Illustrated Flora of the Northern United States and Canada. Vol. 1: Ferns to Buckwheat.* Reprint of 2nd ed. Dover Publications, New York: 92, 93, 94, 96, 97, 98, 101, 158.

Jeanne R. Janish: From *Vascular Plants of the Pacific Northwest* (Hitchcock et al. 1961, 1964 & 1969). University of Washington Publications in Biology, vol. 17. University of Washington Press, Seattle: 47, 54, 58, 67, 74, 106, 107, 109, 116, 117, 128, 131, 132, 135, 145, 171, 176, 192, 196.

Shayna La-Belle Beadman: 13, 15, 17, 21, 22, 26, 27, 28, 29, 33, 37, 39, 40, 41, 45, 49, 59, 60, 64, 69, 70, 72, 73, 77, 80, 82, 83, 88, 91, 95, 96, 100, 101, 103, 114, 117, 127, 154, 191, 194, 200, 203, 204, 206, 207, 210, 211, 212, 213, 215, 216, 217, 218, 219, 220, 221, 222.

Steve Newmaster: 20, 22, 23, 28, 35, 49, 55, 61, 62, 63, 98, 102, 105, 116, 117, 118, 119, 122, 129, 133, 134, 137, 140, 146, 148, 154, 156, 157, 159, 191, 195, 203, 204, 205, 206, 209, 210, 211, 213, 217.

Erika North: 13, 20, 24, 47, 49, 54, 55, 79, 86, 87, 90, 92, 94, 97, 99, 105, 106, 107, 111, 112, 115, 116, 117, 118, 119, 122, 126, 128, 129, 133, 134, 138, 139, 141, 144, 146, 147, 150, 153, 154, 155, 156, 157, 160, 161, 163, 164, 165, 166, 167, 168, 170, 171, 172, 173, 174, 175, 176, 178, 179, 180, 181, 182, 183, 184, 186, 187, 189, 211, 214.

Other Illustrators: **J. Bowles**: 32, 44, 46, 196, 197, 206, 208, 216; **Linda Kershaw**: 61, 87, 105, 150, 206, 207, 208, 211, 215; **Daniel Klassen**: 106, 108, 116, 117, 137, 140, 142, 201; **Emma Thurley**: 13, 56, 71, 81, 99, 120, 132, 199.

INDEX
French Names

Only primary species are listed.

Common Names

Primary species and their main treatments are listed in boldface type.

Scientific Names

Primary species and their main treatments are listed in boldface type.